T0348559

MULTIPLE CRITERIA ANALYSIS FOR AGRICULTURAL DECISIONS

Second Edition

Developments in Agricultural Economics 11

MULTIPLE CRITERIA ANALYSIS FOR AGRICULTURAL DECISIONS

Second Edition

Carlos Romero
Department of Forest Economics and Management,
Forestry School, Technical University of Madrid, Spain

and

Tahir Rehman
School of Agriculture, Policy and Development,
Faculty of Life Sciences, University of Reading, UK

2003

ELSEVIER
Amsterdam · Boston · London · New York · Oxford · Paris
San Diego · San Francisco · Singapore · Sydney · Tokyo

ELSEVIER SCIENCE B.V.
Sara Burgerhartstraat 25
P.O. Box 211, 1000 AE Amsterdam, The Netherlands

© 2003 Elsevier Science B.V. All rights reserved.

This work is protected under copyright by Elsevier Science, and the following terms and conditions apply to its use:

Photocopying
Single photocopies of single chapters may be made for personal use as allowed by national copyright laws. Permission of the Publisher and payment of a fee is required for all other photocopying, including multiple or systematic copying, copying for advertising or promotional purposes, resale, and all forms of document delivery. Special rates are available for educational institutions that wish to make photocopies for non-profit educational classroom use.

Permissions may be sought directly from Elsevier's Science & Technology Rights Department in Oxford, UK: phone: (+44) 1865 843830, fax: (+44) 1865 853333, e-mail: permissions elsevier.com.You may also complete your request on-line via the Elsevier Science homepage (http://www.elsevier.com), by selecting 'Customer Support' and then 'Obtaining Permissions'.

In the USA, users may clear permissions and make payments through the Copyright Clearance Center, Inc., 222 Rosewood Drive, Danvers, MA 01923, USA; phone: (+1) (978) 7508400, fax: (+1) (978) 7504744, and in the UK through the Copyright Licensing Agency Rapid Clearance Service (CLARCS), 90 Tottenham Court Road, London W1P 0LP, UK; phone: (+44) 207 631 5555; fax: (+44) 207 631 5500. Other countries may have a local reprographic rights agency for payments.

Derivative Works
Tables of contents may be reproduced for internal circulation, but permission of Elsevier Science is required for external resale or distribution of such material.
Permission of the Publisher is required for all other derivative works, including compilations and translations.

Electronic Storage or Usage
Permission of the Publisher is required to store or use electronically any material contained in this work, including any chapter or part of a chapter.

Except as outlined above, no part of this work may be reproduced, stored in a retrieval system or transmitted in any form or by any means, electronic, mechanical, photocopying, recording or otherwise, without prior written permission of the Publisher.
Address permissions requests to: Elsevier's Science & Technology Rights Department, at the phone, fax and e-mail addresses noted above.

Notice
No responsibility is assumed by the Publisher for any injury and/or damage to persons or property as a matter of products liability, negligence or otherwise, or from any use or operation of any methods, products, instructions or ideas contained in the material herein. Because of rapid advances in the medical sciences, in particular, independent verification of diagnoses and drug dosages should be made.

First edition 1989 (Volume 5)
Second edition 2003

British Library Cataloguing in Publication Data

Romero, Carlos, 1946-
    Multiple criteria analysis for agricultural decisions. –
    1.Agriculture – Economic aspects – Mathematical models
    2.Multiple criteria decision making
    I.Title II.Rehman, T. (Tahir)
    338.1'015118

    ISBN 0444503439

Library of Congress Cataloging in Publication Data
A catalog record from the Library of Congress has been applied for.

ISBN: 0-444-50343-9
ISSN: 0926-5589 (Series)

♾ The paper used in this publication meets the requirements of ANSI/NISO Z39.48-1992 (Permanence of Paper).

Printed and bound in the United Kingdom
Transferred to Digital Printing, 2011

For our boys Ben, Carlitos and Joe

Contents

Preface   xi

Acknowledgements   xiii

Part one: Multiple criteria in agricultural decisions

1   Main features of the multiple criteria decision-making paradigm   3
     Criticism of the traditional paradigm for decision-making   3
     Economic versus technological decisions   4
     Multiple objectives and goals in agricultural economics   7
     Historical origins of the MCDM paradigm   8
     Plan of the book   10
     Suggestions for further reading   14

2   Some basic concepts   15
     Attributes, objectives and goals   15
     Distinction between goals and constraints   16
     Pareto optimality   17
     Trade-offs between decision-making criteria   18
     A first approximation of the main MCDM approaches   19
     Suggestions for further reading   20

Part two: Multiple criteria decision-making techniques

3   Goal programming   23
     Introductory example for handling multiple criteria in a farm planning model   23
     The role of deviational variables in goal programming   26
     Lexicographic goal programming   27
     Sensitivity analysis in LGP   30
     The graphical method for solving an LGP problem   31
     The sequential linear method for LGP   33
     A brief comment on other LGP algorithms   36
     Weighted goal programming   37
     A critical assessment of goal programming   38
     Some extensions of goal programming   41
     Suggestions for further reading   45

4 **Multiobjective programming**   47

An approximation of the multiobjective programming problem   47
The pay-off matrix in MOP   51
The constraint method   52
The weighting method   54
The noninferior set estimation method (NISE)   55
Multigoal programming   58
Some issues related to the use of MOP techniques   59
Suggestions for further reading   60

5 **Compromise programming**   63

An intuitive treatment of the concept of distance measures   63
A discrete approximation of the best-compromise solution   66
Compromise programming – a continuous setting   68
The method of the displaced ideal   71
Pros and cons of GP, MOP and CP   74
Relationships between different MCDM approaches   75
Suggestions for further reading   78

6 **The interactive multiple criteria decision-making approach**   79

Structure of an interactive MCDM process   80
The STEM method   81
The Zionts and Wallenius method   88
Interactive multiple goal programming   92
An assessment of interactive MCDM approaches   97
Suggestions for further reading   100

7 **Risk and uncertainty and the multiple criteria decision-making techniques** 103

Risk programming techniques in agricultural planning within an
  MCDM framework   103
Compromise-risk programming   108
Game theory models and the MCDM framework   110
Games with multiple goals and goal programming   113
Compromise games   116
Suggestions for further reading   118

Part three: Case studies

8 A compromise programming model for the agrarian reform programme in
  Andalusia, Spain   123
    Background   124
    Trade-off curves for seasonal labour, employment and gross margin   126
    Compromise sets   129
    An approximation of the efficient set in a three-dimensional
        objectives space   132
    Concluding comments   133

9 Livestock ration formulation and multiple criteria decision-making techniques 135
    A livestock ration formulation example   137
    Ration formulation as a WGP problem   138
    Ration formulation as an LGP problem   142
    Ration formulation as an MOP problem   146

10 Livestock ration formulation via goal programming with penalty functions   149
    Penalty functions in diet formulation   149
    Diet formulation as a WGP model with penalty functions   152
    Diet formulation as an LGP model with penalty functions   157
    An assessment   160

11 Optimum fertiliser use via goal programming with penalty functions   163

References   169

Index   185

Preface

The traditional mathematical programming approach for modelling agricultural decision-making processes rests on some fundamental assumptions, relating to both the decision situation being modelled and the decision maker. One of these assumptions is that the decision maker (DM) is seeking to optimise a well-defined single objective. There is now a considerable body of literature that undermines the tenability of this assumption. In reality the DM is usually looking for an optimal compromise between several objectives, many of which are in conflict, or else he is endeavouring to achieve satisficing levels of the goals that he has set for himself. This seems true of farmers everywhere; for instance, a subsistence farmer may be interested in securing adequate food supplies for the family, maximising cash income, increasing the amount of time spent on meeting social obligations and leisure activities, or in avoiding risk and so on. Likewise, a farmer in the developed world may typically be concerned with maximising gross margins, minimising indebtedness, increasing his net worth, striving to be the leading farmer in the area, etc.

Despite the recognition accorded to the existence of multiple objectives in agricultural decision-making, very little seems to have been done by agricultural economists to develop and use methodologies that help model the real-life decision situation. It is an intriguing state of affairs, when one notices that an impressive amount of intellectual effort has been devoted to the development and use of multiple criteria decision-making (MCDM) techniques in management science, water resources research and forest planning. Until the appearance of the first edition of this book, there has hardly been a textbook that introduces agricultural economists to MCDM techniques. It has filled an important gap and continues to satisfy a demand that is growing within the profession.

The book is divided into three parts. The first part, comprising of two chapters, is philosophical in nature and deals with the rationale behind the use of MCDM techniques in decision-making and the fundamental concepts that must be understood to appreciate the nature of these methods. The second part is the largest and contains five chapters, each dealing with the logical structure of a specific MCDM technique and how it is used to model a particular decision problem. Some prominent applications of MCDM techniques to agricultural decision-making are presented in the last four chapters of the third part. The book has been designed for use both as a textbook and as a source of reference; each chapter concludes by suggesting further reading and the extensive list of references at the end of the book should serve the purposes of agricultural economists embarking upon research in this arena. Anyone with a basic knowledge of linear algebra and linear programming can obtain an understanding of the conceptual bases of most of the MCDM techniques presented here.

A comment on the use of the Simplex method of solving problems of constrained optimisation is in order in the context of the use and the development of the MCDM models. All the techniques presented here use the Simplex method, or some variant of it, to solve multiple objectives problems. This fact should not be allowed to confuse the MCDM approaches with the traditional linear programming approach to modelling of decision-making. The Simplex method is used simply as a solution algorithm, just as one would use differentiation for solving problems in calculus. The use of a common solution method does not imply a similarity in the conceptual foundations, purposes and the orientation of the underlying analytical approaches of the single and multiple criteria decision-making models. It should also be pointed out that in writing this book the authors have been concerned primarily with the exposition of the new methods and the explanation of the concepts; and, therefore, the examples for illustration have been selected for their suitability to a particular method and not just for the accuracy and/or the realistic nature of the data being used.

Acknowledgements

This book owes it origin to the collaboration between the two authors that first started between Córdoba University, Spain and The University of Reading, England. This contact, which later developed into a full-scale research programme on multiple criteria decision-making techniques, was initiated back in 1982/83 when the first author was a Visiting Professor with the Farm Management Unit of The University of Reading. This long-term cooperation has been sustained through financial support from The British Council, The University of Reading, Consejería de Educación y Ciencia de la Junta de Andalucía, Comisión Interministerial de Ciencia y Tecnología and, particularly, the Anglo Spanish Joint Research Programme (Acciones Integradas). The authors are grateful to all for their help.

Several colleagues at Córdoba, Madrid and Reading have given us their time generously and the constructive criticisms received have helped to improve the quality of our work. Some of them, Dr. Amador and Mr. Barco from Business School (ETEA), Córdoba, Dr Domingo from Córdoba University and Dra. Mínguez from Technical University of Madrid, have in fact contributed to papers written jointly and we are grateful to them for allowing us to use some of the material published elsewhere. Professor Ken Thomson from Aberdeen University, Scotland was very kind in reading the first drafts of some of the chapters. We are grateful to him for his critical vetting of our ideas. All the inadequacies of this book however remain our responsibility.

We have also been very fortunate in being able to draw upon the secretarial assistance of Mrs Liz Townson in typing the first edition from scratch to create the Word files for us to work on for producing the second edition. In the process she has come to learn the idiosyncrasies of the second author's handwriting.

Mark Meredith and his associates, Helen Dighton and Glynn Seeds, from Waygoose Designers have designed the layout of the second edition of our book expertly. Likewise, the editorial skills of Heather Addison have improved the readability of our text. We are grateful to them all.

We would like to record our thanks to the Editors of the following academic journals for allowing us to draw upon the papers published in those journals:

Journal of Agricultural Economics
American Journal of Agricultural Economics
Agricultural Systems
Journal of the Operational Research Society

Finally, most authors seem to owe their intellectual development to a teacher, a colleague or a mentor. In our case, for Tahir Rehman such a person has been late Harold Casey who always had time for discussing new ideas and a kind word of encouragement tempered by caution, no matter how naïve the initial thought. He had retired due to ill health when we started the 'MCDM Project', but would still come to the Department and in fact helped us to revise the first three papers that we published on the MCDM techniques. Similarly for Carlos Romero, particularly in the early stages of his career, such an influence was Professor Enrique Ballestero, Technical University of Madrid , to whom he owes a great deal.

Carlos Romero   Madrid
Tahir Rehman   Reading
August 2002

Part one
Multiple criteria in agricultural decisions

This part considers the fundamental role that multiple criteria play in agricultural decision-making and then states why the traditional mathematical programming paradigm, particularly linear programming, is inadequate for modelling such decisions; then the concepts underlying the logical structure of the Multiple Criteria Decision Making (MCDM) paradigm are developed to facilitate the understanding of the material in the rest of the book.

Chapter one

Main features of the multiple criteria decision-making paradigm

This introductory chapter criticises the main assumptions around which the structure of the traditional paradigm for managerial decision-making has been built. Two main types of decision-making situations are identified: first, the problems that involve a single decision criterion or objective; and second, situations where several conflicting objectives have to be considered and reconciled. It is argued that decision makers in reality pursue several objectives and therefore the single-criterion traditional paradigm is inadequate for dealing with such situations. A review of the research in agricultural economics reveals that multiple objectives are the rule rather than the exception in agricultural decision-making, whether the decision maker is a farmer or a policy maker. Finally, the origin and historical evolution of the multiple criteria decision-making paradigm is described briefly and the structure of the remainder of the book is outlined.

Criticism of the traditional paradigm for decision-making

The traditional framework (or paradigm in the Kuhnian sense) for analysing decision-making presupposes the existence of three elements: a decision maker (an individual, or a group recognised as a single entity); an array of feasible choices; and, a well-defined criterion of choice, such as utility or profit. The given criterion is then used to associate a number with each alternative so that the feasible set can be ranked, or ordered, to find the optimal value that is attainable for the criterion of choice.

This paradigm has served its purpose well so far. For instance, in consumer theory in determining the equilibrium point or optimum decision of a consumer, the first step is to establish the set of attainable baskets of goods without violating the budget constraint. The utility criterion is then used to measure the possible contribution of each basket before using any commonly used optimisation techniques to identify that basket of goods that maximises the decision maker's (DM) utility. The use of mathematical programming for decision-making shares the same theoretical construct. The feasible solutions satisfy the constraints of the problem, which are later ordered according to a given criterion, or the objective function representing the preferences of the DM. The optimum solution, that is, the highest possible value for the objective function, is found from the feasible set using some mathematical procedure, such as the simplex algorithm.

Notwithstanding the fact that this paradigm is logically sound, it does not reflect the real-life decision situations faithfully. The DM is usually not interested in ordering the feasible set according to just a single criterion but would rather find an optimal compromise involving several objectives. As an illustration, consider decision-making in a large corporation interested not just in maximising profits but in optimising some other objectives such as sales or the growth of the firm. Similarly a firm may aim to achieve some goals that have been fixed *ex ante* for some of its objectives rather than optimise any generally stated objective. Precise conceptual and mathematical distinctions between goals and objectives will be established in the next chapter.

Examples of multiple criteria in decision-making abound. In managing a natural resource like fisheries, a balance among a set of conflicting objectives such as cost, income, sustainability of fish population, etc. must be established. In designing a car engine the problem may be to optimise the conflicting objectives of cost, horsepower and the fuel consumption rate. Likewise, in making a capital budgeting decision the DM is interested not only in maximising the net present value of the portfolio of investments but also in achieving a certain rate of growth in the sales of the company's products, maintaining a certain level of employment and so on.

The situation in agricultural decision-making is no different from the above and the existence of multiple objectives is the rule rather than the exception. A subsistence farmer may be interested in maximising cash income, security of food supplies, increasing leisure, avoiding risk and so on. The commercial farmer may want to maximise gross margin, minimise indebtedness, acquire more land, reduce fixed costs, enhance social standing and so on. Similarly a policy maker, in formulating the policy instruments for bringing about socially desirable allocation of land, may have to consider how the private motive of profit maximisation can be reconciled with the objective of minimising environmental damage. Such examples are common in real life and one has to agree with Zeleny (1982) when he asserts that 'multiple objectives are all around us'.

Another drawback of the traditional paradigm is the assumption that the constraints that define the feasible set are so rigid that they cannot be violated under any circumstances. This is not always the case; in many situations is possible to accept a certain amount of violation of at least some of the constraints. This is true specially in formulating livestock rations or in choosing fertiliser combinations, where the technical knowledge is not precise enough to impose rigid constraints. The preceding discussion should emphasise that the DM whose rationality is suitably explained by the traditional paradigm is an abstract entity whose assumed behaviour is not observed commonly in the real world.

Economic versus technological decisions

It is possible to distinguish between economic and technological problems of choice depending upon whether or not multiple criteria are involved. Friedman (1962) points out that decision problems involving a single-choice criterion should be regarded as technological problems as it is only when multiple criteria have to be considered that decision-making becomes an economic problem. Similarly, Zeleny (1982, chap. 1) argues that logically technological problems consist only of the processes of search and measurement, which can be undertaken using simple tools or very sophisticated methods. But in the technological problem strictly speaking there is no decision-making as one is only searching and measuring. The real decision-making problem arises only when several criteria determine the optimum decision. The following scheme clarifies the distinction between these two concepts.

	Several Criteria	*Single Criterion*
Scarce Means	Economic Problem	Technological Problem
Non-scarce means	No problem ('Nirvana')	

Consider visiting a supermarket for 'choosing' the cheapest bottle of wine. Strictly speaking, it is not a decision-making problem but a technological one, which can be solved easily by a simple procedure of searching based on a comparison of the price/volume ratio. Likewise finding the cropping pattern that maximises the gross margin is also a technological problem involving the search among the feasible cropping patterns, as the DM does not really choose but only searches.

On the contrary, to choose the cheapest bottle of wine, with the highest alcohol content, oldest vintage, and a desirable place of origin and so on is an economic problem. Now it is necessary to find a compromise among several conflicting objectives. The solution has to respect the preferences of the DM with respect to these objectives. Similarly, finding the cropping pattern that maximises gross margin, and minimises risk and indebtedness, is an economic problem whose optimum solution will change from one farmer to another depending on individual preferences.

To illustrate the above ideas, consider the case of a planning agency responsible for developing a small rural region of 1,000 ha of arable land, where it is possible to grow only two crops, A and B. The water requirements of the two crops during the peak season are estimated as 4,000 m^3/ha and 5,000 m^3/ha respectively, and the availability of water during the peak season is 4,200,000 m^3. For the rotational reasons the area of crop B must be less than or equal to the area under A. Allowing X_1 and X_2 to represent areas under crops A and B, the feasible or attainable set is given by:

$$
\begin{aligned}
X_1 + X_2 &\leq 1,000 \\
4000X_1 + 5000X_2 &\leq 4,200,000 \\
-X_1 + X_2 &\leq 0
\end{aligned}
\tag{1.1}
$$

The feasible-set of solutions is represented by the region OABC of Figure 1.1. Assuming now that the preferences of the policy maker are adequately represented by the value-added criterion, and that £1,000/ha and £3,000/ha are contributed by crops A and B respectively, the feasible set can be ordered by the following family of 'iso-value added' lines:

$$1000x_1 + 3000x_2 = V$$

It is easily seen from Figure 1.1 that point A is the feasible point where the value added reaches a maximum level associated with growing 466.66 ha each of crop A and B, giving a value added of £1,866,640.

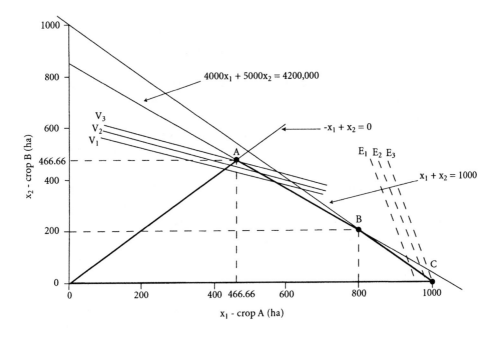

Figure 1.1 An agricultural planning problem with two objectives

As the maximum value added point was picked from the feasible solutions by a process of search and measurement, strictly this is a technological problem and, not an economic one. Now suppose that the policy maker establishes his preferences not only according to the value added, but also by considering the level of employment and that crops A and B require 500 hours/ha and 200 hours/ha respectively; then the feasible set should be ordered using the following 'iso-employment' lines:

$$500x_1 + 200x_2 = E$$

According to this new criterion the optimum solution is given by point C in Figure 1.1 by growing 1000 ha of crop A and nothing of crop B, providing a maximum employment of 500,000 hours.

In this new version of our problem the traditional paradigm can only inform us that if the policy maker wishes to order the feasible set according to both the value added and employment criteria, then his equilibrium point lies along the domain ABC. But to discover the optimum point along this domain, the search techniques are not sufficient and new tools of analysis are needed. In other words, we must analyse that economic problem within the structure of a different paradigm. This is the main purpose of this book; thus, the conceptual and operational features of two paradigmatic structures, the goal and multiobjective programming approaches, are being presented for analysing decision-making problems that involve multiple and conflicting objectives and goals.

Multiple objectives and goals in agricultural economics

The need to find a balance among multiple objectives and goals in agricultural planning is now well established. In a seminal piece of research Ruth Gasson (1973) asked 100 farmers in Cambridgeshire, England, to rank a set of attributes representing their values for being in business in farming. Sixteen attributes were ranked: at the top of the ranking appears independence or doing the work you like, while at the bottom are job security or belonging to the farming community.

A growing body of literature has succeeded Gasson's research. Some brief comments on this work are in order. Smith and Capstick (1976) interviewed 111 farmers in Northeast-Arkansas and ranked ten goals according to the preferences attached to them. The goals preferred most were to stay in business and stabilise income; and the less preferred ones were increase in net worth and larger farm size. Similarly, Harper and Eastman (1980) established a hierarchy of goals for small farm operators in New Mexico, and Cary and Holmes (1982) studied the actual goals of graziers in South-West Queensland, discovering a wide economic and sociological variety amongst these. For a good survey of such early research see Patrick and Kliebenstein (1980).

Despite this empirical evidence, modellers in agricultural economics have not paid too much attention to the crucial role that should be given to several objectives and goals in building decision-making models. However there are some exceptions to this attitude. Wheeler and Russell (1977) were the first to introduce several goals in a farm level decision-making model in agriculture. They analyse the planning problem of an hypothetical 600 acres mixed farm in the United Kingdom and consider the goals of maximum gross margin, minimum seasonal cash exposure and provision of stable employment for the permanent labour throughout the year. Bartlett and Clawson (1978) tackle a ranch planning problem in Sacramento, considering three goals: red meat production, use of fossil fuel energy and profits. Marten and Sancholuz (1982) have used a framework of multiple goals to analyse ecological land use planning problems in Mexico.

In subsistence agriculture two early examples are of interest. Flinn et al. (1980) analyse a decision-making problem in the Philippines where six goals are taken into account, from the production of enough rice for family subsistence to the generation of sufficient cash surplus. Barnett et al. (1982) use a similar approach to tackle a decision-making problem in Senegal.

At a regional level Hitchens et al. (1978) and Thampapillai and Sinden (1979) examine a land allocation problem in Australia, where two conflicting objectives of money income and environmental benefits are considered. Vedula and Rogers (1981) tackle another land allocation problem in India to achieve the two objectives of maximising net income benefits and total irrigated cropped area.

In the regional planning field, Bazaraa and Bouzaher (1981) introduce several goals in a plan for the agricultural sector in Egypt, and Romero et al. (1987) have used a framework of multiple objectives to examine the implementation of an agrarian reform programme in Andalusia, Spain.

This brief commentary supports the view that the agricultural decision makers, be they farmers or policy makers, have a strong motivation to seek optimisation or satisfaction of several objectives or goals rather than to pursue the maximisation of a single criterion. As a result an important body of literature has been developing where the agricultural decision-making models are formulated under the realistic assumption of multiple objectives and goals.

Historical origins of the MCDM paradigm

The analysis of problems involving the multiple criteria decision-making (MCDM) paradigm has been perhaps the fastest growing area of operational research and management science (OR/MS) during the last 35 years. According to Vincke (1986), in 1975 3.5% of the papers presented to the Congress organised by the Association of European Operational Research Societies (EURO Congress) were devoted to MCDM topics, the percentage increasing to 14% in 1985; that is, these days one out of every seven papers in the EURO Congresses is concerned with some aspect of the MCDM paradigm. Some recent bibliographies on MCDM such as Zeleny (1982) and Stadler (1984) include more than 1,000 and 1,700 references, respectively. Likewise surveys of some specific approaches to MCDM such as goal programming (GP) include more than 900 references Schniederjans (1995). In a recent survey Steuer et al. reveal that in just over five years, between 1987 and 1992, more than 1,200 refereed journal articles were published on MCDM.

The above situation raises two interesting questions. First, what started this 'scientific revolution' in OR/MS? Second, when did the turning point occur, when this 'new' paradigm was accepted by researchers for application to real problems?

The answer to the first question lies in two papers: Koopmans (1951) and Kuhn and Tucker (1951). The first paper has developed the concept of efficient or non-dominated vector, which plays a crucial role in MCDM; while the second one has formulated the multiobjective or vector maximisation problem and the optimality conditions for the existence of non-dominated solutions are derived. Another crucial contribution to the development of the MCDM paradigm is Charnes, Cooper and Ferguson (1955), where in analysing the problem of obtaining 'constrained regressions' estimates for an executive compensation problem they have presented an embryonic form of the goal programming (GP) approach. Some years later, Charnes and Cooper (1961) presented a more complete formulation of GP in an Appendix to their book *Management Models and Industrial Applications of Linear Programming.*

These pioneering ideas have been taken up by others and gradually developed further. For instance Zadeh (1963) was the first to suggest the weighting method for solving multiple objective programming (MOP) problems, Marglin (1967) proposed the constraint method to solve MOP problems, Geoffrion (1968) established the concept of proper efficient solutions, and Ijiri (1965) made considerable improvements to GP with pre-emptive weights.

The point at which the above approaches may be considered to have matured into an MCDM paradigm is perhaps 1972, as in October that year the first international conference on MCDM was held at the University of South Carolina, USA. More than sixty papers were presented at the conference attended by about 250 participants. The proceedings were later published in a book edited by Cochrane and Zeleny (1973) marking the point of acceptance of the MCDM paradigm as part of 'normal science' in the Kuhnian sense.

The meeting at South Carolina agreed to form a Special Interest Group on MCDM. This group has evolved into the International Society on Multiple Criteria Decision-making, which was formed in 1979; currently this body has more than 1,200 members from 80 different countries. First as the Special Interest Group, and then later as the International Society, this organisation has held biannual international conferences; the last one was held at Semmering (Austria) in February 2002. Table 1.1 provides a list of the edited volumes of these conferences, all of which have been published by Springer-Verlag. Other international groups on MCDM are the EURO Working Group on Multicriteria Decision Aid, formed in 1985, and the Multiobjective Programming and Goal Programming (MOPGP) group, formed in 1994. A sample from the proceedings of the different MCDM organisations is listed in Table 1.2.

As mentioned earlier, since the beginning of the 1970s a real 'explosion' in the number of papers on both the theoretical aspects of the MCDM paradigm and its application published in OR/MS journals has occurred. Nowadays it would be difficult to find an issue of these journals without any paper on the theoretical or practical aspects of MCDM. Certain journals have even published special issues devoted entirely to the topic of MCDM, as listed in Table 1.3.

The undoubted success of the MCDM paradigm has led to the appearance of the *Journal of Multi-Criteria Decision Analysis.* The rationality of a journal specifically devoted to MCDM is perhaps questionable, since to some extent it implies the existence of two distinct decision-making environments: single criterion, and multiple criteria, which amounts to a contradiction of our arguments so far. And as Zeleny (1982, p. 74) says: 'No decision-making occurs unless at least two criteria are present. If only one criterion exists, mere measurement and search suffice for making a choice'. The single criterion decision-making is just an old paradigm that has been superseded by the new MCDM one, and is in fact a particular case of the new MCDM paradigm.

Plan of the book

This book has two primary purposes: first, to demonstrate that a real decision-making environment in agriculture involves several objectives and goals; second, to explain how the different MCDM techniques can be applied to analyse and solve such decision problems.

Chapter 1 examines the inadequacy of the traditional paradigm to model real decision problems, particularly in agriculture. To achieve that goal and in order to appreciate the different techniques to be presented, some basic concepts are introduced in Chapter 2.

The second part of the book is concerned with explaining the logical structure of the most commonly used MCDM techniques and to show how they are used to model decision processes in agriculture. This part is divided into five chapters. Chapter 3 is devoted to GP, where the DM, instead of optimising a single objective, strives to satisfy as much as possible of a set of conflicting goals. Chapter 4 analyses the problem of simultaneous optimisation of several objectives through multiobjective programming as a means of generating the set of efficient solutions. In Chapter 5 compromise programming is studied as a method to choose the best compromise for the DM from among the efficient solutions. In Chapter 6 some interactive techniques are presented, where the preferences of the DM are elicited through an 'interaction' (that is, a conversation) involving the DM, the model and the analyst (modeller). In Chapter 7 agricultural decision-making under conditions of risk and uncertainty is brought within the scope of the MCDM paradigm.

Some applications of MCDM techniques to real decision problems in agriculture are presented in the third part of the book. Thus, Chapter 8 presents a case study where compromise programming is used to resolve the conflict between private and social objectives in the design of an agrarian reform programme. In Chapters 9 and 10 the role of MCDM techniques in ration formulation is discussed, demonstrating the use of penalty functions in formulating nutritionally desirable diets for livestock. In Chapter 11 the use of penalty functions is applied to a real case study of optimum fertiliser mixing for sugar beet.

Table 1.1 Proceedings of the conferences organised by the International Society on MCDM (formerly Special Interest Group on MCDM)

Title of the conference	Editor(s) of proceedings	Place where conference held	Year
Multiple Criteria Decision-making	H. Thiriez and S. Zionts	Jouy-en-Josas (France)	1975
Multiple Criteria Problem Solving	S. Zionts	Buffalo (USA)	1977
Multiple Criteria Decision-making – Theory and Application	G. Fandel and T. Gal	Königswinter (Germany)	1979
Organizations: Multiple Agents with Multiple Criteria	J.N. Morse	Delaware (USA)	1980
Essays and Surveys on Multiple Criteria Decision-making	P. Hansen	Mons (Belgium)	1982
Decision-making with Multiple Objectives	Y.Y. Haimes and V. Chankong	Cleveland (USA)	1984
Multiple Criteria Decision-making – Toward Interactive Intelligent Decision Support Systems	Y. Sawaragi, K. Inoue and H. Nakayama	Kyoto (Japan)	1986
Improving Decision Making in Organizations	A.G. Lockett	Manchester (UK)	1988
Multiple Criteria Decision Making	A. Goicoechea, L. Duckstein and S. Zionts	Fairfax (USA)	1990
Multiple Criteria Decision Making	G.H. Tzeng, H.F. Wang, P.U. Wen and P.L. Yu	Taipei (Taiwan)	1992
Multicriteria Analysis	J. Climacao	Coimbra (Portugal)	1994
Multiple Criteria Decision Making	G. Fandel, T. Gal and T. Hanne	Hagen (Germany)	1996
Trends in MultiCriteria Decision Making	T.J. Stewart and R.C. van den Honert	Cape Town (South Africa)	1997
Research and Practice in Multiple Criteria Decision Making	Y.H. Haimes and R.E. Steuer	Charlottesville (USA)	1998

Table 1.2 A sample of proceedings of other conferences on MCDM

Title of the conference	Editor(s) of proceedings	Place where conference held	Year	Number of papers	Publisher
Multiple Criteria Decision-making	M. Zeleny	Kyoto (Japan)	1975	16	Springer-Verlag, 1976
Conflicting Objectives in Decisions	D.E. Bell, R.L. Keeney and H. Raiffa	Laxenburg (Austria)	1975	18	John Wiley and Sons, 1977
Multiobjective and Stochastic Optimisation	M. Grauer, A. Lewandoski and A.P. Wierzbicki	Laxenburg (Austria)	1981	24	International Institute for Applied Systems Analysis, 1982
Theory and Practice of Multiple Criteria Decision-making	C. Carlsson and Y. Kochetkov	Moscow (Russia)	1981	9	North-Holland, 1983
Macro-Economic Planning with Conflicting Goals	M. Despontin, P. Nijkamp (Belgium) and J. Spronk	Brussels	1982	14	Springer-Verlag, 1984
MCDM: Past Decade and Future Trends	M. Zeleny	Washington (USA)	1982	11	JAI Press Inc., 1984
Interactive Decision Analysis	M. Grauer and A.P. Wierzbicki	Laxenburg (Austria)	1983	28	Springer-Verlag, 1984
Multiple Criteria Decision Methods and Applications	G. Fandel, J. Spronk and B. Matarazzo	Sicily (Italy)	1983	20	Springer-Verlag, 1985
Multicriteria Decision Support	P. Korhonen, A. Lewan-Dowski and J. Wallenius	Helsinki (Finland)	1989	44	Springer-Verlag, 1991
Multi-Objective Programming and Goal Programming	M. Tamiz	Portsmouth (UK)	1994	23	Springer-Verlag, 1994
Advances in Multiple Objective and Goal Programming	R. Caballero, F. Ruiz and R.E. Steuer	Málaga (Spain)	1996	41	Springer Verlag, 1997
Multiple Objective and Goal Programming	T. Trzaskalik and J. Michnik	Ulstron (Poland)	2002	33	Physica-Verlag, 2002

Table 1.3 A sample of special issues of OR/MS journals devoted to MCDM

Journal	Title	Editor(s)	Year/Volume	Number of Papers
Management Science (TIMS Studies in the Management Sciences)	Multiple Criteria Decision-making	M.K. Starr and M. Zeleny	1977/6	15
Computers and Operations Research	Mathematical Programming with Multiple Objectives	M. Zeleny	1980/7	11
Computers and Operations Research	Generalized Goal Programming	J.P. Ignizio	1983/10	10
Regional Science and Urban Economics	Multiobjective Decision Analysis in a Regional Context	P. Nijkamp and P. Rietveld	1983/13	7
Large Scale Systems	Multicriterion Optimisation and Decision Support	M.G. Singh and A.P. Sage	1984/6	6
IIE Transactions	Multiple Criteria Decision-making in Production Planning and Scheduling	K.D. Lawrence and R.E. Steuer	1984/16	4
Management Science	Multiple Criteria Decision-making	J. Spronk and S. Zionts	1984/30	10
European Journal of Operational Research	Multiple Criteria Decision-making	T. Gal and B. Roy	1986/25	12
European Journal of Operational Research	Multicriteria Analysis	J.P. Brans, M. Despontin and P. Vincke	1986/26 and P. Vincke	14
Engineering Costs and Production Economics	Multiple-Criteria Decision-Making	M.T. Tabucanon	1986/10	11
Mathematical Modelling	The Analytic Hierarchy Process	L.G. Vargas and T.L. Saaty	1987/9	25
Naval Research Logistics	Multiple Criteria Decision Making	S. Zionts and M.H. Karwan	1998/35	12
Engineering Costs and Production Economics	Multicriterion Production Systems	M.T. Tabucanon and V. Chankong	1990/20	13
Water Resources Bulletin	Multiple Objective Decision Making in Water Resources	K.W. Hipel	1992/28	18
Agricultural Systems	Multiple Criteria Analysis in Agricultural Systems	T. Rehman and C. Romero	1993/41	11
Computers and Operations Research	Implementing Multiobjective Optimization Models	J.L. Ringuest	1993/19	16

Suggestions for further reading

Walsh (1970) provides an excellent critical presentation of the traditional paradigm of choice involving a single criterion. The evolution of the concept of optimality from a single to a multiple objective environment is clearly explained by Keen (1977) and a superb explanation of the philosophical foundations and the operational aspects of the MCDM paradigm can be seen in Zeleny (1982). Besides Zeleny's book there are now several textbooks covering the theoretical and practical aspects of MCDM. Among them the following can be recommended: Cohon (1978), Goicoechea et al. (1982), Chankong and Haimes (1983), Yu (1985) and Steuer (1986) covering the full range of the MCDM techniques. The books by Lee (1972), Ignizio (1976), Romero (1991) and Schniederjans (1995) are concerned exclusively with GP. Recently Ballestero and Romero (1998) have made an attempt to link the MCDM paradigm with traditional economic analysis.

Patrick and Kliebensteins (1980) is a good reference to support the hypothesis of multiple objectives and goals in agricultural decision-making. A critical survey of MCDM applications in agricultural planning can be seen in Rehman and Romero (1987a). For a review of MCDM techniques within an agricultural environment see Thampapillai (1978). Finally, an extensive survey of natural resources management problems tackled within a MCDM framework can be seen in Romero and Rehman (1987a). A special issue of *Agricultural Systems* (1993) edited by Rehman and Romero demonstrates the applicability of this approach in agriculture and natural resources management.

Chapter two
Some basic concepts

To appreciate what is involved in modelling the agricultural decision processes via MCDM paradigm, certain fundamental concepts related to the MCDM techniques need to be made clear. First, we establish the conceptual differences among attributes, objectives and goals, and draw the distinction between goals and the conventional interpretation of constraints. Second, the idea of an efficient or a Pareto optimal solution is explained as it is essential to the development of the multiobjective programming approach. Third, the idea of trade-offs between objectives and goals is treated as a corollary of the concept of efficient solutions. Finally, a first approximation to the main MCDM techniques is provided. Some of the concepts introduced in this chapter may have the same dictionary meanings, for example, goals and objectives and, in the context of some problems, they can be used interchangeably. However, for the economic problems being analysed within a MCDM framework, nuances of the conceptual differences have to be established as these ideas have a meaning and usefulness only within the theoretical structure within which they have been created.

Attributes, objectives and goals

Attributes can be defined as a decision maker's values related to an 'objective' reality. These values can be measured independently from his desires and in many cases can be expressed as a mathematical function $f(\underline{x})$ of the decision variables. For example, in the simple case presented in the preceding chapter the policy maker establishes his preferences according to two attributes: value added and level of employment. The attributes represent the values of the decision maker (DM) and are measured, in monetary units (pounds sterling) and hours of work, independently of the DM's desires, and are expressed as mathematical functions of the decision variables. In fact, the mathematical equivalents for the attributes value added and level of employment are $1000x_1 + 3000x_2$ and $500x_1 + 200x_2$ respectively. Some of the examples of attributes include gross margin, seasonal cash requirements and indebtedness in a farm planning problem and the intake of crude protein, metabolisable energy in a diet formulation model.

Given this definition of an attribute, an objective represents directions of improvement of one or more of the attributes, implying the sense of either 'the more of the attribute, the better' or 'the less of the attribute, the better'. The first case is a maximisation process whilst the second one is minimisation. Stating an objective, therefore, implies the maximisation or the minimisation of the functions representing one or several attributes reflecting the values of the DM; thus, maximising value added, minimising risk and minimising cost are examples of typical objectives. In general, objectives take the form Max $f(\underline{x})$ or Min $f(\underline{x})$ and they are not attributes even though they are derived from them.

For example, the objective Max $1000x_1 + 3000x_2$ in the example used in Chapter 1 represents maximisation of the value added, but in some cases an objective is derived from more than one attribute. For instance, if a DM considers two attributes represented by the functions $f_1(\underline{x})$ and $f_2(\underline{x})$, then the objective maximisation would be given by:

$$\text{Max } w_1 f_1(\underline{x}) + w_2 f_2(\underline{x})$$

where w_1, w_2 represents the importance attached, respectively, by the DM to each of the attributes.

Before defining a goal let us state what is an aspiration level or a target. A target is an acceptable level of achievement for any one of the attributes. On combining an attribute with a target we have a goal. For instance, in the above example if the policy maker wants a particular cropping pattern to yield a value added of at least £2,000,000 we have a goal, which is expressed as $1000x_1 + 2000x_2 \geq 2,000,000$. In some cases, however, the DM may aim for an exact achievement of the target; for instance, if the DM wants all of the land to be cultivated, then the goal is $x_1 + x_2 = 1,000$. In general, goals take the form $f(\underline{x}) \geq t$ or $f(\underline{x}) \leq t$ or $f(\underline{x}) = t$, where t represents the aspiration level or the target value.

It should be pointed out that a modeller can consider two different types of goals: first, the goals that represent a DM's desires such as requiring the gross margin or the value added of a particular farm plan to reach a specific value; and second, goals that refer to the existence of limited resources, such as water for irrigation, or to the fulfilment of an explicit or implicit constraints, for example crop rotations. In this sense, the goals do not represent a DM's desires in the strictest sense, but only as flexible constraints. Such goals are very helpful in mimicking real life because they relax the complete rigidity of the traditional constraints, as assumed in traditional linear programming models.

In short, then, in a farm planning problem gross margin is an attribute, to maximise gross margin is an objective, and to achieve a gross margin of at least a certain target is a goal. Finally, a criterion encompasses the three preceding concepts: that is, criteria are the attributes, objectives or goals to be considered relevant for a certain decision-making situation. Hence MCDM is a general framework or paradigm involving several attributes, objectives or goals.

Distinction between goals and constraints

At this point the reader may wonder what the actual difference is between goals and constraints. In fact, being inequalities both goals and constraints have identical mathematical structure and thus look exactly the same. The difference between them lies in the meaning that is attached to the right-hand parameter of the inequality; for goals, it is a target aspired by the DM, which may or may not be achieved and when it represents a rigid restraint, it has to be satisfied otherwise the solution will be infeasible. For example, the inequality $1000x_1 + 3000x_2 \geq 2,000,000$ referring to the value-added aspiration being considered in our planning example, could be either a goal or a constraint, depending on how the right-hand side parameter is interpreted. It is a restraint if the inequality must be satisfied under any circumstance, and it a goal if £2,000,000 is treated as a target for the DM.

It follows from this that goals can be considered as soft constraints which can be violated without producing infeasible solutions. The amount of violation can be measured by introducing positive and negative deviational variables. For example, the goal referring to the achievement of a value added of £2,000,000 can be represented by the following equality:

$$1000x_1 + 2000x_2 + n - p = 2,000,000$$

The variables n and p account for deviations from the achievement of a goal from its target. For instance, if $n = $ £500,000 it means that the goal has fallen short by £500,000. So the amount of violation of a goal in the sense of an under-achievement is represented by the negative deviational variable n. The positive deviational variable does the opposite, that is, it indicates the amount by which a goal has surpassed its target. For instance, $p = $ £300,000 means that the goal has exceeded its target by £300,000, so that the value added achieved is £2,300,000. So the amount of violation of a goal in the sense of an over-achievement is represented by the positive deviational variable p. Generally, a goal can be expressed as:

Attribute + Deviational variables = Target

or in mathematical terms as:

$$f(\underline{x}) + n - p = t$$

The deviational variable is a very useful devices for two different reasons: first, it is a simple and interesting way to impart flexibility to constraints, that is, to convert rigid constraints into goals or soft constraints; second, it is the first step to build a goal programming model, which is the most widely used approach within the general MCDM framework, as explained in the next chapter.

Pareto optimality

The concept of Pareto optimality plays a vital role in traditional economic theory and is also a fundamental idea within the MCDM paradigm, as all the approaches within this paradigm look for efficient or Pareto optimal solutions.

The efficient or Pareto optimal solutions are feasible solutions such that no other feasible solution can achieve the same or better performance for all the criteria under consideration and strictly better for at least one criterion. In other words, a Pareto optimal solution is a feasible solution for which an increase in the value of one criterion can only be achieved by degrading the value of at least one other criterion. To clarify this concept, let us consider a hypothetical farm planning problem with the following three feasible solutions for the three different criteria:

Gross margin (£)	Seasonal labour (hours)	Indebtedness (£)
200,000	500	50,000
200,000	600	50,000
300,000	700	60,000

Assuming that the DM wants the gross margin to be as large as possible and wishes both the use of seasonal labour and the level of indebtedness to be as small as possible, then the second solution is clearly non-efficient, since it offers the same gross margin and indebtedness as the first one, but requires more seasonal labour; thus, the second solution will never be chosen by a rational DM. The third solution, however, is Pareto optimal. In fact, it has a larger requirement in indebtedness and for seasonal labour but it also offers a bigger gross margin. To choose between the first and the third solutions is an economic problem, where a real decision must be taken according to the preferences of the DM for each one of the three attributes considered. All the MCDM techniques aim to obtain solutions that are efficient in the Paretian sense as defined above.

Trade-offs between decision-making criteria

The concept of a Pareto optimal solution leads to another crucial concept in MCDM: the value of trade-off between two criteria; that is the amount of achievement of one criterion that must be sacrificed to gain a unitary increase in the other one. So if we have two efficient solutions x^1 and x^2 the trade-off value between the jth and kth criteria would be given by:

$$T_{jk} = \frac{f_j(\underline{x}_1) - f_j(\underline{x}_2)}{f_k(\underline{x}_1) - f_k(\underline{x}_2)}$$

where $f_j(\underline{x})$ and $f_k(\underline{x})$ represent the two objective functions being considered; thus, in our example the trade-off value between gross margin and seasonal labour for the first and the third solutions is:

$$T_{12} = \frac{300,000 - 200,000}{700 - 500} = 500$$

The trade-off T_{12} indicates that each hour of decrease of seasonal labour use implies a decrease of £500 of gross margin; that is, the opportunity cost of one hour of seasonal labour is £500 of gross margin. The trade-off T_{13} between gross margin and indebtedness and the trade-off T_{23} between seasonal labour and indebtedness would be given by:

$$T_{13} = \frac{300,000 - 200,000}{60,000 - 50,000} = 10$$

$$T_{23} = \frac{700 - 500}{60,000 - 50,000} = 0.02$$

that is, the opportunity cost of increasing the indebtedness by £1 is £10 of gross margin or 0.02 hours of seasonal labour.

Besides being a good index for measuring the opportunity cost of one criterion in terms of another under consideration, these trade-off values also play a key role in the analysis of interactive techniques as presented in Chapter 6.

A first approximation of the main MCDM approaches

The above distinctions between attributes, objectives and goals allow us to give a first approximation to the main MCDM approaches. Thus, if the DM must take a decision within an environment of multiple goals the method to use is goal programming (GP), which is accomplished by minimising the deviations from the desired levels of targets through the addition of positive and negative deviational variables permitting either the under- or over-achievement of each goal, as explained in the next chapter.

When an environment of multiple objectives is involved, then multiobjective programming (MOP) is used, where an efficient set of solutions is generated first before separating the Pareto optimal feasible solutions from the non Pareto-optimal ones. Next an optimum compromise for the DM from among the efficient solutions is sought, respecting the preferences of the decision maker.

Finally, if the environment within which the DM must take his decision is characterised by several attributes, the approach to be considered is multi-attribute utility theory (MAUT). The purpose of MAUT is to build a utility function with a number of arguments equivalent to the number of attributes under consideration. The MAUT approach is usually applied to decision problems with a discrete number of feasible solutions; it is not considered in this book because its real possibilities of application in agricultural decision-making, where the choice sets are not discrete but continuous, are very limited. The very strong assumptions about the preferences of DM that are required for the implementation of MAUT is another factor that restricts its application.

Suggestions for further reading

Generally, in the literature on decision-making within a single-criterion framework, no distinction is usually made between attributes, objectives and goals. The term attribute is not used, and the terms goal and objective are used interchangeably. Until recently even in the MCDM literature these terms have been used indistinguishably, giving the erroneous impression that they are the same. However, the conceptual distinctions made in this chapter are necessary to clarify the different approaches within the MCDM framework, and are advocated strongly by leading researchers in the field. For lucid discussions of these issues see Zionts (1980, pp. 540–541), Ignizio (1982, pp. 26–27) and Zeleny (1982, pp. 14–19 and 225–228) among others. Eilon (1972) has explained the relationship between goals and constraints excellently, in a very readable paper.

A reader interested in exploring the multi-attribute utility theory can consult Keeney and Raiffa (1976), which is a classic reference on the topic. Similarly, a more formalised treatment of value trade-offs can be seen in Chankong and Haimes (1983, pp. 331–336).

Part two
Multiple criteria decision-making techniques

This part is concerned with the exposition of the logical structures of the most commonly used MCDM techniques and to show how they are used to model the decision process. There are five chapters, each dealing with a specific technique or approach.

Chapter three
Goal programming

This chapter deals with goal programming (GP). It is perhaps the oldest MCDM technique and its general aim is a simultaneous optimisation of several goals, by minimising the deviations from the desired targets for each of the objectives and what is actually achievable in relation to the targets set. The minimisation process can be accomplished by several methods, each being a specific variant of GP. In this chapter, however, we deal primarily with the two best-known and most widely used such variants: lexicographic goal programming (LGP) and weighted goal programming (WGP). LGP accomplishes the minimisation process by attaching pre-emptive or absolute weights to the sets of goals situated in different priorities, that is, the fulfilment of a set of goals situated in a certain priority is immeasurably preferable to the achievement of any other set placed in a lower priority. Hence in LGP the higher priority goals are satisfied first, and it is only then that the lower priorities are considered. The WGP variant, on the other hand, considers all goals simultaneously within a composite objective function comprising the sum of all the respective deviations of the goals from their aspiration levels. The deviations are weighted according to the relative importance of each goal. In short, in WGP the relative importance of the goals is dealt with using their relative weights, while in LGP the absolute goals are handled by their rankings.

This chapter has four broad purposes. First, it shows why the traditional linear programming (LP) model is generally not suitable for dealing with multiple criteria. Second, the conceptual and logical structure of GP and its main variants, LGP and WGP; are explained. Third, modelling of problems involving multiple criteria through the use of GP is demonstrated. Finally, the pitfalls associated with the use of GP are pointed out.

Introductory example for handling multiple criteria in a farm planning model

In this section a hypothetical farm planning situation is presented to illustrate the inadequacies inherent in the LP model for dealing with multiple-criteria decision-making. This example is used later to analyse various aspects of GP in subsequent sections.

The situation to be modelled is the case of a typical Spanish farmer in Lerida county wishing to invest in irrigation systems for part of his land to set up an orchard of pears and peaches. The planning data are given in Table 3.1. It is assumed that the working capital available during the first year is £15,000, while for the second, third and fourth years it is limited to £7,000 per annum. Annual availability of pruning labour is limited to 4,000 hours, while the labour available for harvesting both species is 2,000 hours per annum. A maximum of 1,000 own tractor hours are available during any year for tillage. Finally, the two crops are harvested during different periods.

Table 3.1 Planning data for hypothetical farm planning problem

Decision variables		Pear trees (ha) x1	Peach trees (ha) x2
Net present value of investment in trees (£/ha)		6250	5000
Resource requirements			
Working capital (£/ha) -	Year 1	550	400
	Year 2	200	175
	Year 3	300	250
	Year 4	325	200
Annual labour (man-hours/ha) -	Pruning	120	180
	Harvesting	400	450
Machinery for tillage (hours/ha)		35	35

The objectives of the farmer are: (a) to maximise the net present value (NVP) of investment in the plantation; (b) to minimise the borrowing for the working capital over the next four years; (c) to minimise the hiring of casual labour for pruning and harvesting; and (d) to minimise the use of the contracted tractor hours. This is clearly a situation when maximising NVP can be in conflict with minimising the dependence on the borrowed money, the hired labour and the contracted use of tractors.

The inclusion of the first two objectives is obvious and requires no clarification. As regards the other two objectives, they represent the assumption that the farmer does not wish to incur the effort of obtaining and organising casual labourers or hiring tractor services. Although there were no significant differences between the wages for casual and permanent labour or the cost of using own or hired tractors, one of the goals of the farmer is to do all the jobs on the farm using permanent labourers and his own tractors.

One could solve this problem as an ordinary LP model, by first treating one of the objectives on its own, such as NPV, and then maximising it. The other objectives can be considered as constraints along with those that define the availability of resources. In the constraints for the cash resource the possibility of transferring the surplus in one year to the next has been included.

Thus, if for the first year there is a surplus of cash, this excess will be equal to:

$$15,000 - 550x_1 - 400x_2$$

The second year the cash available will be the £7,000 already there plus the above surplus. Therefore, the actual inequality securing a financial equilibrium during the second year will be:

$$200x_1 + 175x_2 \leq 7,000 + 15,000 - 550x_1 - 400x_2$$

Manipulating the above inequality we obtain:

$$750x_1 + 575x_2 \leq 22,000$$

The cash flow constraints of the second and third year are obtained in a similar way.
The problem then is:

$$\text{Max } z = f(x_1, x_2) = 6250\, x_1 + 500\, x_2$$

subject to

$$
\begin{aligned}
500\, x_1 + 400\, x_2 &\leq 15,000 \\
570\, x_1 + 575\, x_2 &\leq 22,000 \\
1,050\, x_1 + 825\, x_2 &\leq 29,000 \\
1,375\, x_1 + 1,025\, x_2 &\leq 36,000 \\
400\, x_1 &\leq 2,000 \\
450\, x_2 &\leq 2,000 \\
35\, x_1 + 35\, x_2 &\leq 1,000 \\
\text{and} \quad x_1, x_2 &\geq 0
\end{aligned}
\tag{3.1}
$$

and it has the solution x_1 (pear trees) = 5 ha and x_2 (peach trees) = 4.44 ha with an NPV of £53,450. The harvesting labour has been used completely, while other resources have not been.

How should the decision maker receive this solution? It is doubtful if this solution will be acceptable to him as it yields rather low NPV and leaves considerable amounts of various resources unused. However, this strategy is chosen as the optimal one by LP because: (1) the objectives formulated as restraints are satisfied first before maximising NPV; and (2) each feasible solution must satisfy the constraints imposed on the solution space exactly. This approach where a single objective is optimised while treating others as restraints can produce disappointing solutions. For instance, in our example, if borrowing is minimised by considering other objectives as restraints (including the generation of a minimum NPV of £60,000), there is no feasible solution.

This example should not be taken to imply that this method is meaningless. In fact, through parametric variations of the right-hand side of the objectives expressed as constraints, it is possible to generate efficient or non-inferior solutions as explained in the next chapter. However, we must emphasise that dealing with several objectives by introducing them as fixed right-hand side values of an LP problem can work in specific instances, but it is not satisfactory as a general approach to multiple-criteria decision-making. In the following sections the above problem is developed as a GP model and then compared with the results given by the above LP formulation.

The role of deviational variables in goal programming

In setting up the GP model the set of inequalities (3.1) are treated as goals, g_i, instead of constraints. The right-hand side elements are targets, which may or may not be achieved. For each goal, two associated variables, n and p, called the deviational variables are introduced that convert inequalities into equalities, so that:

$$
\begin{aligned}
6250x_1 + 5000x_2 + n_1 - p_1 &= 200{,}000 \text{ that is } g_1 \\
550x_1 + 400x_2 + n_2 - p_2 &= 15{,}000 \text{ that is } g_2 \\
750x_1 + 575x_2 + n_3 - p_3 &= 22{,}000 \text{ that is } g_3 \\
1050x_1 + 825x_2 + n_4 - p_4 &= 29{,}000 \text{ that is } g_4 \\
1375x_1 + 1025x_2 + n_5 - p_5 &= 36{,}000 \text{ that is } g_5 \\
120x_1 + 180x_2 + n_6 - p_6 &= 4{,}000 \text{ that is } g_6 \\
400x_1 \quad\quad + n_7 - p_7 &= 2{,}000 \text{ that is } g_7 \\
450x_2 + n_8 - p_8 &= 2{,}000 \text{ that is } g_8 \\
35x_1 + 35x_2 + n_9 - p_9 &= 1{,}000 \text{ that is } g_9
\end{aligned}
\tag{3.2}
$$

In order to point out that what the DM really wants is to maximise NPV, an artificially high target of £200,000 for NPV has been set, which is impossible to be achieved given the resources assumed in our example.

The deviational variables account for deviations from the aspiration level set for the achievement of a goal. For instance, if $n_1 = £50{,}000$, it means that g_1 has fallen short by £50,000. In other words, the actual attainment of g_1 is £150,000. So under-achievement of a goal is represented by a negative deviational variable.

The positive deviational variable does the opposite job, that is, it indicates the amount by which a goal's achievement has surpassed its aspiration level. For instance, $p_9 = 100$ means that goal g_9 has surpassed its target by 100 hours; that is, the number of tractor hours required is 1100. So positive deviational variables represent over-achievement of goals.

A goal cannot be both under-achieved and over-achieved. Hence, in a solution at least one of the deviational variables for each goal is zero. When a goal g_i matches its aspiration level exactly then $n_i = p_i = 0$. If a certain goal's achievement must be greater than or equal to its target then its negative deviational variable is unwanted and it has to be minimised. If a certain goal must be less than or equal to its target, then the positive deviational variable is unwanted and it has to be minimised. Finally, if a certain goal must be exactly equal to its target, then both positive and negative deviational variables are unwanted and they have to be minimised.

The general purpose of GP is to minimise the unwanted deviational variables. That minimisation process can be undertaken in different ways; among them, those most widely used in practice are (1) to attach absolute or pre-emptive weights to the unwanted deviational variables and (2) to attach relative or non pre-emptive weights to the unwanted deviational variables. We now explain both of these approaches in turn.

Lexicographic goal programming

This approach (LGP) was first introduced by Charnes and Cooper (1961, pp.756–757) and developed further by Ijiri (1965), Lee (1972) and Ignizio (1976). It is assumed that a decision maker (DM) can explicitly define all the goals that are relevant to a particular planning situation. Further, LGP assumes that a DM can not only attach priorities to these goals, but does so in a pre-emptive fashion. In other words, the fulfilment of the goals in a specific priority, Q_i, is immeasurably preferable to the fulfilment of any other set of goals situated in a lower priority, Q_j. Many authors refer to this situation using the notation $Q_i >>> Q_j$. In LGP, higher priority goals are satisfied first and it is only then that lower priorities are considered; hence, the lexicographic order.

To illustrate the structure of LGP, assume that in our example the DM's priority Q_1 is made up of goals g_2, g_3, g_4, and g_5. That is, for the DM the first goal that must be satisfied in an absolute and pre-emptive way is the one which assumes the equilibrium between the outflows of cash and the financial resources available, permitting the transfer of funds from the periods of surplus to the ones with a deficit. The first component to minimise in the lexicographic process will be given by $[p_2 + p_3 + p_4 + p_5]$. The next priority in order of importance, Q_2, is made up of goal g_9, which refers to the use of machinery for tillage. Thus, the second component is given by the positive deviation variable, p_9. The priority Q_3 is made up of goal g_1, referring to the maximisation of the NPV, thus giving the third component, n_1. Finally, the last priority, Q_4, is made up of goals g_6, g_7, and g_8, referring to the minimisation of hired casual labour for pruning and harvesting. Thus, the last component of the minimisation process is given by $[p_6 + p_7 + p_8]$. The whole lexicographic minimisation problem is then:

$$\text{Min } \underline{a} = [(p_2 + p_3 + p_4 + p_5), (p_9), (n_1), (p_6 + p_7 + p_8)] \tag{3.3}$$

This vector, called the achievement function, replaces the objective function in the conventional LP model. Each component of this vector represents the deviation variables (positive or negative) that must be minimised in order to make sure that the goals ranked in this priority come closest to the expected achievement levels.

In general, the achievement function is stated as:

$$\text{Minimise } \underline{a} = [h_1\,(\underline{n}, \underline{p}),\, h_2\,(\underline{n}, \underline{p}),\, \ldots,\, h_1\,(\underline{n}, \underline{p})]$$

or alternatively,

$$\text{Minimise } \underline{a} = [a_1,\, a_2,\, \ldots,\, a_1]$$
where $a_1 = h_1\,(\underline{n}, \underline{p})$ is a function of the deviational variables and   (3.4)

We seek to find the lexicographic minimum of $\underline{a}$; that is, the minimisation of the vector (3.4) implies the ordered minimisation of its components. It is therefore necessary to find, first, the smallest value of the first component a_1, then the smallest value of the next component, a_2, and so on.

Several of the research papers published on LGP write the achievement as:

$$\text{Minimise } Z = [Q_1 h_1\,(\underline{n}, \underline{p}) + Q_2 h_2\,(\underline{n}, \underline{p}) + \ldots + Q_1 h_1\,(\underline{n}, \underline{p})]$$ (3.5)

where Q_1 denotes the first priority with an infinitely larger weight than the priority Q_2, and so on. Both Zeleny (1982, p. 223 and p. 299) and Ignizio (1985, pp. 30-31) have pointed out, quite correctly, that to express the achievement function this way is misleading as the summation in (3.5) reduces the expression to a scalar, which is contrary to what is meant to be conveyed. This expression in (3.5) is not only meaningless it also leads to incorrect developments of LGP as well as wrong algorithms for solving such problems. In our discussion and development of the LGP models, the achievement function is as specified correctly in the expression (3.4) is used.

By combining the achievement function of (3.3.) with the set of goals in (3.2), we obtain the following LGP model for our example.

$$\text{Min } \underline{a} = [(p_2 + p_3 + p_4 + p_5),\, (p_9),\, (n_1),\, (p_6 + p_7 + p_8)]$$ (3.6)

subject to

$$
\begin{aligned}
6250x_1 + 5000x_2 + n_1 - p_1 &= 200{,}000 \text{ that is } g_1 \\
550x_1 + 400x_2 + n_2 + p_2 &= 15{,}000 \text{ that is } g_2 \\
750x_1 + 575x_2 + n_3 - p_3 &= 22{,}000 \text{ that is } g_3 \\
1050x_1 + 825x_2 + n_4 - p_4 &= 29{,}000 \text{ that is } g_4
\end{aligned}
$$

$$1375x_1 + 1025x_2 + n_5 - p_5 = 36{,}000 \text{ that is } g_5$$
$$120x_1 + 180x_2 + n_6 - p_6 = 4{,}000 \text{ that is } g_6$$
$$400x_1 \qquad\quad + n_7 - p_7 = 2{,}000 \text{ that is } g_7$$
$$450x_2 + n_8 - p_8 = 2{,}000 \text{ that is } g_8$$
$$35x_1 + 35x_2 + n_9 - p_9 = 1{,}000 \text{ that is } g_9$$

$$x \geq 0, n_j \geq 0, p_j \geq 0 \qquad\qquad i = 1, 2 \text{ and } j = 1, \ldots, 9$$

This LGP can be solved by any of the many possible algorithms. Their explanation is deferred to the next sections. However, using one of them, we obtain the following optimum solution:

$$x_1 = 19.18 \qquad\qquad x_2 = 9.38$$
$$n_1 = £33{,}250 \qquad\quad p_1 = 0$$
$$n_2 = £699 \qquad\qquad p_2 = 0$$
$$n_3 = £2{,}221 \qquad\quad p_3 = 0$$
$$n_4 = £1{,}122 \qquad\quad p_4 = 0$$
$$n_5 = n_6 = 0 \qquad\quad p_5 = p_6 = 0$$
$$n_7 = 0 \qquad\qquad\quad p_7 = 5{,}672 \text{ hours}$$
$$n_8 = 0 \qquad\qquad\quad p_8 = 2{,}221 \text{ hours}$$
$$n_9 = 0 \qquad\qquad\quad p_9 = 0$$

This solution permits complete achievement of the goals that make up the first two priorities. As for goal g_1 that represents the third priority and sets the target for NPV of at least £200,000, it was not reached, producing a negative deviation of £33,250 or an NPV of £166,750. Finally, with respect to goals g_6, g_7 and g_8 that constitute the last priority Q_4, only the goal g6 specifying the non-use of casual labour for pruning was completely satisfied. The goal g_7 has a positive deviation of 5,672 hours of casual labour for harvesting pear trees. The goal g_8 also has a positive deviation of 2,221 hours of casual labour required for harvesting peach trees.

What would be the attitude of a DM to this solution? The rational DMs are likely to prefer this LGP solution to the one given by LP as it yields a large NPV (£113,300) and each resource is utilised completely. A possible disadvantage is that 7,893 hours of casual labour are being hired during the harvesting period, which can generate organisational problems considering the objectives of the DM as stated already. Anyway, a trade-off of NPV worth £113,300 against the effort of organising 7,893 hours of casual labour seems to be very profitable for the farm. It should also be pointed out that if the casual labour wage is higher than the equivalent permanent labour charges, it will be necessary to correct the expectation on NPV provided by the LGP model.

Further, in formulating the achievement function, the DM may attach weighting factors to goals within a predetermined priority if necessary. Thus, in our example, if the DM believes that in priority Q_4, goal g_6 is twice as important as goals g_7 and g_8 (that is, the use of casual labour for pruning is twice as important as the use of casual labour for harvesting), the last component of the achievement function would be $(2p_6 + p_7 + p_8)$.

Sensitivity analysis in LGP

One possible weakness of the LGP approach lies on the great amount of information required from the decision maker (DM), as he has to provide the analyst with information related to targets, weights attached to each goal placed within a certain priority and pre-emptive ordering of preferences. When the DM is not very confident of the values of any of these parameters, the implementation of an appropriate sensitivity analysis is especially recommended. This kind of analysis allows an easy exploration of the several effects of rearrangement of priorities and of setting different values for targets on alternative planning strategies.

In our example (see Table 3.2) we have explored the effects of rearrangement of priorities on alternative planning strategies. There are four priority levels, with the possible rearrangements of priorities being 4! = 24. Of these permutations only six solutions are different and are given below for comparison.

Table 3.2 Sensitivity analysis in lexicographic goal programming

Solutions	Decision variables			Deviations from targets		
	x_1 (ha)	x_2 (ha)	NPV (£)	Casual labour (hrs)	Contracted tractor (hrs)	Cash deficit (£)
I	19.18	9.38	33,250	7,893	0	0
II	5	4.44	146,550	0	0	0
III	0	35.12	24,400	16,125	229	0
IV	28.57	0	21,437	9,428	0	3,284
V	0	40	0	19,200	400	5,000
VI	32	0	0	10,800	120	8,000

Solution I is optimal only when the order of the first two priorities is reversed. Solution II is optimal for twelve of the twenty-four rearrangements or priorities and solution III is obtained only when the third priority is moved to the second place. Solution IV is obtained when the second priority is moved to the first place and the third priority is moved to the first place and the first priority is moved to the second place. Finally, solution VI is obtained in the four cases where the third priority is moved to the first place and the second priority is moved to the second or fourth place.

On making some changes to the targets or aspiration levels of the various goals, the following effects can be observed.

1 If the target g_1 is lowered as far as to £166,775 the optimum solution does not change, but if it goes below this amount the NPV is worsened and the deviations from g_6, g_7 and g_8 decrease. For example, if the target for g_9 is set to £150,000 the optimum solution is x_1 = 19 ha, x_2 = 6.25 ha with a deviation of 6,712 hours of casual labour during the harvesting period.

2 If the aspiration level for g_9 is lowered the optimum solution changes, reducing the NPV. On the contrary, if the aspiration level for g_9 is set higher (up to 1,229 hours) the optimum solution changes and improves the NPV. For example, if the target for g_9 is set to 1,050 hours the optimum solution is x_1 = 15 ha, x_2 = 15 ha, with deviations of £31,250 in the NPV, 500 hours of pruning casual labour and 8,750 hours of harvest casual labour.

3 If the targets for g_6, g_7 and g_8 change (increase or decrease) the values of the optimum decision variables do not change. The only changes are in the deviation variables for g_6, g_7 and g_8.

In the example above, the sensitivity analysis has been implemented by following a brute force approach. However, analysis can be carried out in a very efficient way just by using the optimal or last tableau when some LGP algorithms, such as the modified simplex method, are used (see Suggestions for further reading); thus it is not necessary to reformulate and resolve the problem from the beginning each time that a change in value of one of the parameters is implemented.

Although the explanation of techniques like the modified simplex method is beyond the scope of this chapter, the readers interested in this topic could consult Ignizio (1976 chap. 4, 1982 chap. 19), where the following cases are studied: (1) a discrete change in the value of the weights associated with the deviational variables; (2) a discrete change in the value of the targets; (3) addition of a new goal; (4) reordering of priority levels and other variations on the theme.

The lack of commercial software packages for sensitivity and parametric analysis in LGP limits the possibilities of actual application of these techniques considerably. In fact, although these methods do not require too many calculations they are very difficult to implement manually, even for problems of moderate size.

The graphical method for solving an LGP problem

The graphical method for solving LGP models can be used only when two decision variables are involved, therefore it has a very limited practical application. However, it is useful in explaining the functioning of the LGP technique itself and also for appreciating the algebraic algorithms used for solving larger LGP models.

The graphical method is an adaptation of the diagrammatic approach for solving LP problems with two variables. We start by plotting all the goals as straight lines (Figure 3.1) in two dimensions; and the decision variables are represented by the two axes. The effect of an increase in the value of any of the deviational variables is shown by arrows. The deviational variables to be minimised are circled[1]. It should be pointed out that as the region bounded by the goal g_5 becomes more restricted than the region bounded by the goals g_2, g_3 and g_4, only the straight line of goal g_5 is included in the graph; in other words, goals g_2, g_3 and g_4 are redundant.

As the minimisation process under consideration is lexicographic the goals placed in the highest priority g_2, g_3 and g_4 are considered first. The goal g_5 is satisfied (and therefore goals g_2, g_3 and g_4 are also satisfied) when the deviational variable p_2 is minimised. The shaded area OAB ($p_2 = p_3 = p_4 = p_5 = 0$; $x_1 \geq 0$, $x_2 \geq 0$) of Figure 3.1 represents the set of alternative optimum solutions for the goals placed in the first priority. In other words, each point within or in the boundary of the triangle OAB permits the complete satisfaction of the four goals making the first priority (no borrowing for working capital over the next four years).

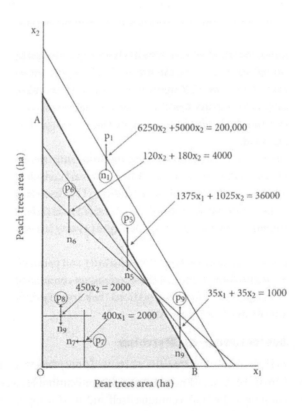

Figure 3.1 Graphical solution for the first priority

The next lower priority considers only the goal g_9. To satisfy that goal the deviational variable p_9 must be minimised. The new shaded area OABC of Figure 3.2 represents the set of alternative optimum solution for the first two priorities.

The third priority considers only the goal g_9. To satisfy that goal we must minimise the deviational variable n_1, but as can be seen from Figure 3.2 it is not possible to make $n_1 = 0$ without degrading the goals situated in the higher priorities. It is easy to verify that the minimum value of n_1, without degrading the higher priorities, corresponds to point B. Therefore, the coordinates of this point represent the optimum solution for the first three priorities and moreover as there are no alternative optimum solutions (i.e. the best solution for goal g_1 is given just by a single point instead of a straight line) then point B represents the optimum solution for the whole problem. In fact, there is no sense in trying to satisfy the goals g_6, g_7 and g_8 making up the last priority because to achieve that purpose it would be necessary to minimise p_6, p_7 and p_8, which is impossible without degrading the goals situated in higher priorities. Therefore, the goals situated in the last priority are redundant and have not played any role in the order of priorities considered in our example. We will return to this matter of redundant goals in the last section.

The sequential linear method for LGP

The main advantage of the sequential linear method (SLM) with respect to other methods for solving LGP problems is that it only requires the conventional simplex method of solution used in LP. The SLM solves a sequence of LP problems in a manner similar to what was done with the graphical method.

The first LP problem of the sequence minimises the first component of the achievement vector, subject to the constraints (equalities) corresponding to priority Q_1. The second linear program minimises the second component of the achievement vector, subject to the constraints corresponding to priorities Q_1 and Q_2, as well as the values of the deviational variables in priority Q_1 which were found in the preceding solution. This sequential procedure continues until the last linear program is solved or until in one of the problems of the sequence there are no alternative optimum solutions.

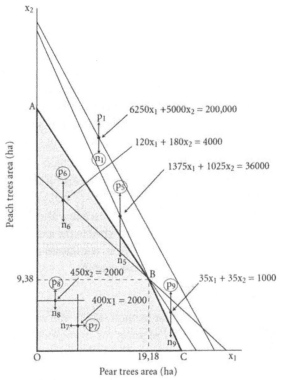

Figure 3.2 Graphical solution for the second priority

On applying this approach to our problem we obtain the following sequence of LP problems.

Problem 1 – the first priority level

Minimise $a_1 = p_2 + p_3 + p_4 + p_5$

subject to

$$550x_1 + 400x_2 + n_2 - p_2 = 15,000$$
$$750x_1 + 575x_2 + n_3 - p_3 = 22,000$$
$$1050x_1 + 825x_2 + n_4 - p_4 = 29,000$$
$$1375x_1 + 1025x_2 + n_5 - p_5 = 36,000$$

There are alternative optimum solutions[2] for decision variables and $p_2 = p_3 = p_4 = p_5$ $=0$[2]. In fact, the alternative optimum solutions correspond to the domain OAB of Figure 3.1.

Actually the first iteration both for the SLM and the graphical method does exactly the same job.

Problem 2 – the second priority level

In this iteration the second component of the achievement function is minimised subject to the goals corresponding to priorities Q_1 and Q_2 and substituting p_2, p_3, p_4 and p_5 by the optimum values obtained in the first iteration (i.e. $p_2 = p_3 = p_4 = p_5 = 0$) giving the following LP problem:

Minimise $a_2 = p_9$

subject to

$$
\begin{aligned}
550x_1 + 400x_2 + n_2 &= 15{,}000 \\
750x_1 + 575x_2 + n_3 &= 22{,}000 \\
1050x_1 + 825x_2 + n_4 &= 29{,}000 \\
1375x_1 + 1025x_2 + n_5 &= 36{,}000 \\
35x_1 + 35x_2 + n_9 - n_9 &= 1{,}000
\end{aligned}
$$

Again there are alternative optima for decision variables and $p_9 = 0$, corresponding to the domain OABC of Figure 3.2 and once again the equivalence between the SLM and the graphical method is apparent.

Problem 3 – the third priority level

Pursuing the logic of the SLM further, in this iteration the third component of the achievement function is minimised, subject to the goal constraints corresponding to the second problem, setting $p_9 = 0$ and augmenting that structure with the goal constraint which makes up priority Q_3 resulting in the following LP problem:

Minimise $a_3 = n_1$

subject to

$$
\begin{aligned}
550x_1 + 400x_2 + n_2 &= 15{,}000 \\
750x_1 + 575x_2 + n_3 &= 22{,}000 \\
1050x_1 + 825x_2 + n_4 &= 29{,}000 \\
1375x_1 + 1025x_2 + n_5 &= 36{,}000 \\
35x_1 + 35x_2 + n_9 &= 1{,}000 \\
6250x_1 + 5000x_2 + n_1 - p_9 &= 200{,}000
\end{aligned}
$$

Now the optimum solution is $x_1 = 19.18$ ha, $x_2 = 9.38$ ha, $n_1 = £33,250$, corresponding to the point B in Figure 3.2. With this result we could go to the next priority level, setting up the last problem, but as multiple optima do not exist for problem 3, the present solution is optimal with respect to all priorities[3]. All we do is to substitute the optimum values of decision variables into the goals g_6, g_7 and g_8 in order to obtain the values for the deviational variables.

The SLM has been applied extensively in practice, mainly because of the availability of efficient LP software. But it should be noticed that a straightforward application of this approach demands a large number of calculations. In fact, to apply the SLM method involves solving as many LP problems as the number of priorities, l, of the model (where for the first $l-1$ problems there are alternative optimum solutions). Some refinements of the SLM have been proposed to mitigate this situation, where the solutions of different LP problems are linked in a single computer run by an external routine, at least by using the MPSX package (see Ignizio and Perlis 1979).

Another possible weakness of the SLM lies in its inability to deal with sensitivity and post-optimality analysis using efficient methods, as pointed out in the previous section. For that reason, when a detailed sensitivity analysis is to be undertaken it is better to use other LGP methods such as the ones mentioned below.

A brief comment on other algorithms for LGP

Algorithms other than the sequential linear programming (SLM) method do exist for solving LGP problems and they work in a more efficient way than the rather crude approach of the SLM method.

The first modification to the SLM is basically a straightforward extension of the 'two phase' simplex procedure for solving LP problems and it permits multiple instead of only two phases. A detailed explanation of the 'modified or multiphase simplex' with computer codes written in FORTRAN can be found in Lee (1972, chaps 5 and 6) and Ignizio (1976 chap. 3 and Appendix). This algorithmic approach is very convenient when a sensitivity and post-optimality analysis is implemented. In fact, such an analysis can be achieved in a very efficient way from the last or optimal scenario obtained by the modified or multiphase simplex (see Ignizio 1976, chap. 4).

Arthur and Ravindran (1978, 1980a) have proposed a 'partitioning' algorithm for LGP problems, which can be considered an improved and refined version of the SLM method. They have tested a FORTRAN code for this algorithm with several problems of different sizes and they claim that their algorithm takes as little as 12% and never more than 60% of the computer time of the codes proposed by Lee and Ignizio based on the modified or multiphase simplex.

Schniederjan and Kwak (1982) have presented another algorithm for LGP based on Baumol's simplex method known as the multi-step simplex method. It requires a fewer number of tableau elements per iteration than the modified or multiphase simplex. A

FORTRAN code for this algorithm can be seen in Schniederjans (1984, App A and B). A comparative analysis of the relative advantages of the four computer codes commented on above can be seen in Olson (1984).

It should also be mentioned that algorithms for solving LGP problems with integer variables are also available. These algorithms can cope with complete or partial integerisation of variables and with variables that have to assume values of either zero or one (Arthur and Ravindran 1980b; Garrod and Moores 1978; Ignizio 1976, chap. 5; Lee and Morris 1977). However, for dealing with non-linear relationships, both LGP and LP seem to suffer from similar difficulties; there are only a few LGP algorithms that can deal with non-linearity. There are two methods of dealing with non-linearities in GP that should be mentioned. One is an adaptation of the Griffit-Stewart's linear approximations approach and the other consists of the modified pattern search proposed by Hooke and Jeeves, as described in Ignizio (1976, chap. 6 and Appendix) along with the appropriate computer codes in FORTRAN.

Weighted goal programming

This variant of goal programming (GP) considers all the goals simultaneously in a composite objective function, which minimises the sum of all the deviations among the goals from their aspiration levels. The deviations are weighted according to the relative importance attached to each goal by the decision maker (DM).

For expository purposes let us take the goals g_2, g_3, g_4 and g_5 from the above example as the rigid constraints; that is, the aspiration levels of these goals must be satisfied under all circumstances to obtain a feasible solution. In other words, we are assuming now that the DM is not ready to allow any kind of violations of goals g_2, g_3, g_4 and g_5 (i.e. he does not want to increase his indebtedness). Thus we propose to build a WGP with five goals (g_1, g_2, g_3, g_4 and g_5) and four rigid constraints.

First of all it must be pointed out that the target for the NPV has been fixed as £175,600 because this is the maximum NPV compatible with the four rigid cash flow constraints. To fix that target at £200,000, as was done in the LGP formulation, is not correct for the WGP case because it would mean penalising the deviations over the maximum achievable, which does not make any sense.

The variables of the objective function must represent percentage deviations from the targets rather than absolute deviations because of the widely different units of measurements used for the goals g_1, g_6, g_7, g_8 and g_9. The target for g_1 is measured in pounds sterling, the one for g_9 is given in tractor-hours, while for g_6, g_7 and g_8 the targets are expressed in man-hours. Moreover, the absolute values of the various targets are very different. Under these conditions the sum of the absolute deviations from the goals is meaningless and, what is more important, the solution provided by the model can be biased because more importance is given (something akin to an artificial extra weight) to the goal for NPV, as its absolute deviation is much higher than the absolute deviations from targets of the other goals. To avoid these problems, the model minimises the sum of the percentage deviations from targets. Thus the WGP formulation of our problem is:

$$\text{Minimise } z = w_1 \frac{n_1}{175600} \cdot \frac{100}{1} + w_2 \frac{p_2}{4000} \cdot \frac{100}{1} + w_3 \frac{p_7}{2000} \cdot \frac{100}{1} + w_4 \frac{p_8}{2000} \cdot \frac{100}{1} + w_5 \frac{p_9}{1000} \cdot \frac{100}{1}$$

subject to

$$
\begin{aligned}
550x_1 + 400x_2 &\leq 15{,}000 \\
750x_1 + 575x_2 &\leq 22{,}000 \\
1050x_1 + 825x_2 &\leq 29{,}000 \\
1375x_1 + 1025x_2 &\leq 36{,}000 \\
6250x_1 + 5000x_2 + n_1 - p_1 &= 175{,}000 \\
120x_1 + 180x_2 + n_6 - p_6 &= 4{,}000 \\
400x_1 + n_7 - p_7 &= 2{,}000 \\
450x_2 + n_8 - p_8 &= 2{,}000 \\
35x_1 + 35x_2 + n_9 - p_9 &= 1{,}000
\end{aligned}
$$

(3.7)

$$x_1, x_2 \geq 0 \qquad\qquad n_j, p_j \geq 0 \qquad\qquad j = 1 \text{ and } j = 6, \ldots, 9$$

where $w_1, \ldots, w_5$ represent the weights attached to the deviational variables. Mathematically, this is an orthodox LP problem and therefore requires no extension of the simplex algorithm. Different solutions can be obtained by attaching different values to these parameters. For instance, if $w_1 = w_2 = \ldots, = w_5 = 1$ then the solution is the same as the one provided earlier. Assume now that the farmer attaches greater importance to earnings than to the reduction in hiring casual labour or to contract tractor-hours. In that case higher weight should be attached to the deviational variables. The LP solution does not change if the weight attached to n_1 is increased up to 6.5 times the values associated with other deviational variables. But beyond that the optimal solution changes to $x_1 = 22.87$ ha and $x_2 = 4.44$ ha.

A critical assessment of goal programming

Although GP is a very attractive method for combining the logic of optimisation in LP with the DM's desire to satisfy several goals, it is not without drawbacks. This is especially true when GP is applied mechanically without being aware of the logic underlying the approach. In this section we point out certain situations where using GP techniques for decision-making will either produce unexpected results or they will be inappropriate. There are basically five such situations: (1) the possibility of identical solutions provided by the conventional LP and the GP models for a given problem; (2) the inherent assumption of LGP that even though trade-off between goals can take place within a given priority but they cannot be traded off across the boundaries of different priorities; (3) susceptibility of

GP to produce optimal solutions that are inferior as defined in Chapter 2; (4) the theoretical problem with LGP where the maximisation of its achievement function is not the same as optimising the utility function of a decision maker; and (5) the practical problem inherent in LGP when the number of priorities is excessive that can lead to a naïve prioritisation.

When the optimum achievement function takes the form:

$$\underline{a}^* = [0, 0, \ldots, a_r, a_s, \ldots, a_t]$$

and there are no alternative optimum solutions for the rth problem of the sequence, then the optimum solution obtained is the same as the one for an LP model which optimises the goal considered in priority Q_r as the objective function, and sets the goals in the first $r-1$ priorities as constraints. So in our example the LGP solution given earlier can also be obtained by maximising the NPV subject to the cash and contracted tractor constraints. This problem will arise if the targets of the goals considered in the first $r-1$ priorities have been set too pessimistically and the target of the goal considered in the priority Q_r too optimistically. This equivalence of solutions can occur in a WGP model when the aspiration level of one objective is set too high (that is, solution is infeasible) and the aspiration levels of the other objectives too low (that is, it is very easy to achieve).

Because of the fact that both GP and LP formulations of a given problem can yield identical solutions, under certain circumstances, the analysts using GP techniques can conclude, quite erroneously in our view, that either GP is superfluous or of limited usefulness. This is a misleading observation as the equivalence of solutions is to do with the formulation of the problem rather than the nature or potential usefulness of GP. That kind of drawback is present in some real applications of GP to agricultural planning problems (Flinn et al. 1980; Barnett et al. 1982) as has been pointed out elsewhere (Romero and Rehman 1983).

The peculiar way the possibility of trade-offs among a decision maker's goals is treated by LGP deserves attention before its application in real-life problems. In this context the idea of a trade-off implies by how much the achievement of a goal, say g_1, will have to be sacrificed for a unitary increase in another goal, g_2, as compensation. In lexicographic structures of GP models, the trade-off between goals is possible only when they are in the same priority. This possibility is not allowed across different priorities as they are assumed to be independent of each other in a pre-emptive way. This appears to make the LGP model rather restricted but in fact this situation is not very different from the conventional LP structure where no trade-off is assumed to exist between the objective function and the restraint set[4]. However, in practical applications of LGP when a decision maker is not confident about the pre-emptive ordering of priorities, sensitivity analysis of the final solution should be given greater significance than is normally accorded to this activity.

Zeleny and Cochrane (1973, pp. 377–383) were the first to note that in applying GP to situations where the targets for several goals have been set at very pessimistic levels, it is possible to generate an optimal solution that is dominated by other feasible solution(s) in the sense explained in Chapter 2 of this book. The possibility of a dominated solution is likely when the optimal solution of a GP model includes zero values for a relatively large number of deviational variables. In this situation the first remedy is to conduct a parametric analysis of the aspiration levels assumed in the model. This analysis should indicate whether or not it is possible to increase the satisfaction of some goals without reducing the achievement of others. Another approach is to use the test recommended by Hannan (1980) to verify whether a GP solution is efficient or not. If the test is negative, then Hannan suggests the use of a method to generate the set of non-dominated GP solutions. Masud and Hwang (1981) have also proposed a method that guarantees efficient GP solutions. Recently Tamiz and Jones (1996) have provided a state-of-the-art review dealing with the testing of GP solutions for efficiency and also on restoring efficiency of solutions if inefficient solutions are being generated by GP.

Gerard Debreu (1959, pp. 72–73) proved that the lexicographic orderings of preferences are inconsistent with the utility function structure underlying these preferences; that is, a lexicographic ordering cannot be represented by a utility function. Hence, as pointed out by several authors (e.g. Harrald et al. 1978), the achievement function of the LGP models does not optimise the utility function of the decision maker. The analysts who use GP techniques differ on the significance of this feature. To some it is a serious weakness of GP, while others argue that lexicographic ordering is a better representation of a DM's decision environment than what can be achieved by maximising his utility. However, the real significance of this debate may only be theoretical. From a practical point of view it is perhaps a matter of the attitude that an analyst adopts towards the representation of a decision maker's environment and his objectives.

Finally, let us examine the naïve prioritisation problem. All the algorithms devised to solve LGP problems assume that the first problem of the sequence has alternative optimal solutions. When there are no alternative optimal solutions (multiple ties) the algorithm stops and avoids considering the goals belonging to lower priorities. Thus, if the jth problem of the sequence has no alternative optimal solution, the goals situated in priorities lower than the jth one are redundant.

When the number of priorities used to model a given problem is not large (say, between three and five) then the likelihood of multiple ties in the penultimate priority could be high. The goals situated in the last priority can therefore still play an active role in the outcome of the LGP model. However, when the number of priorities is large there is a high likelihood that goals situated in the lower priorities will be treated as redundant and consequently ignored.

An excessive number of priorities can imply that the model is not reflecting the actual wishes of the DM and some of the goals may only be ornamental. Such a prioritisation is naïve and has to be avoided. This weakness is especially serious when the size of the problem is small in relation to the number of priorities. Thus, an LGP model with about ten pre-emptive priorities, some thirty decision variables and fifty constraints (a frequently occurring size in the literature on LGP) can lead to a situation where every goal, except the goals situated in the first three or four priorities, is redundant.

Thus, in our example, goals g_6, g_7 and g_8 making up the fourth priority are redundant and do not play any role in the optimisation process, because there were no alternative optima in the penultimate priority. In other words, an excessive prioritisation of goals leads to an unrealistic model. We would therefore advocate dividing the goals into a small number of pre-emptive priorities (Ignizio 1976, p. 182, suggests five as an upper bound) and attaching non pre-emptive weights to the different goals according to their relative importance to the DM. A detailed empirical approach to the analysis of redundancy in GP can be seen in Amador and Romero (1989).

Some extensions of goal programming

This chapter has provided an expository analysis and assessment of the two main variants of goal programming (LGP and WGP), describing the possibilities of their application in agricultural planning. However, the variants discussed are not the only available methods for the GP approach. Since the early 1970s, several major methodological extensions have been made to GP. We wish to point out the main features of some of these extensions as they appear to be particularly suited for dealing with problems encountered in building agricultural planning models.

In many planning models some goals (for example, the financial structure of a farm) must be introduced as ratios or as fractional goals. Dealing with such problems leads to what is called fractional GP. Since its introduction by Kornbluth (1973) for financial planning, the fractional GP problem has not been solved completely satisfactorily; there are, however, some useful algorithmic approaches that can be used in this context (Kornbluth and Steuer 1981).

Some authors suggest that in certain problems, instead of a pre-emptive or non pre-emptive minimisation of the sum of deviational variables, it would be more useful to minimise the maximum of deviations. This extension of GP, which is a closely related to the compromise programming approach, as discussed in Chapter 5, has been labelled as MINIMAX GP (Flavell 1976). When the achievement of all their goals must be greater than or equal to their targets, the mathematical structure of a MINIMAX GP can be represented below:

Minimise d

subject to

$$n_j \leq d$$
$$f_j(\underline{x}) + n_j - p_j = t \tag{3.8}$$
$$\underline{x} \in F$$

where d is the maximum deviation, $f_j(\underline{x})$ is the mathematical representation of the jth attribute, t_j is the target for the jth goal and F is the feasible set.

In order to introduce elements of risk associated with the achievement of goals, the chance constrained programming technique devised by Charnes and Cooper (1959) can be successfully incorporated into the GP models. Keown (1978) was the first to show how this can be done in a straightforward fashion.

In all the GP formulations that have been commented upon so far, there is the underlying assumption that any deviation with respect to the target is penalised according to a constant marginal penalty, no matter how large the deviation. In fact, the weight attached to a deviational variable is the measure of penalty incurred as a result of deviation from a given target. Graphically that assumption means that the penalty is the slope of a unique straight line as shown in Figure 3.3 for the case where the negative deviational variable is minimised, in Figure 3.4 for the opposite case (i.e. the positive deviational variable is minimised), and in Figure 3.5 for the case where both variables are minimised. The slopes of these straight lines are the weights attached to the deviational variables in the achievement function of a LGP structure or in the objective function of a WGP model.

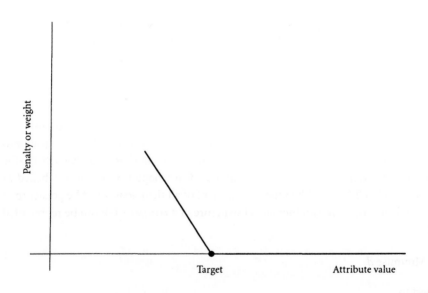

Figure 3.3 A goal's achievement greater than or equal to its target

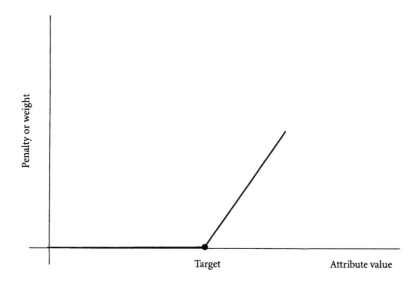

Figure 3.4 A goal's achievement less than or equal to its target

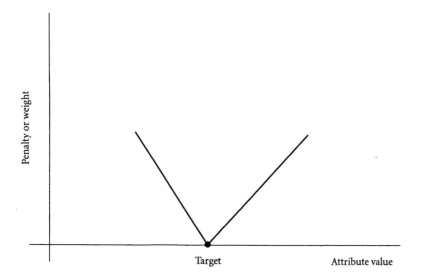

Figure 3.5 A goal's achievement exactly equal to its target

Kvanli (1980) was the first to suggest that in some problems it would be more realistic to consider different marginal penalties according to the magnitude of the deviational variable. This gives rise to GP with penalty functions. Figure 3.6 illustrates a very simple case where the GP model has a three-sided penalty function. In fact, the deviations up to $t - a$ units are penalised according to a marginal penalty V_1, while for the deviations larger than $t - a$ units and smaller than $t - b$ units the marginal penalty is $1/2$. For deviations larger than $t - 1$ units there is an infinite penalty what is tantamount to setting a rigid constraint, indicating that the attribute considered must not be less than b units. GP with penalty functions are explained in some detail in Chapters 10 and 11 with reference to diet ration formulation and optimum fertiliser use problems. Recently Romero (1991, chap. 6), and Jones and Tamiz (1995) show several refinements of GP models with penalty functions.

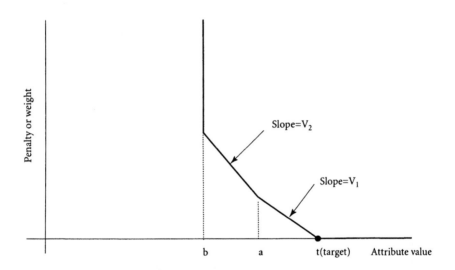

Figure 3.6 Three-sided penalty function

All the above extensions to the basic GP model when taken together have been termed as generalised GP by Ignizio (1983). In fact, this general framework should apply to any MCDM technique where targets have been assigned to all the objectives in the sense of the Simonian concept of 'satisficing'. Ignizio utilises a DM's desire to satisfy several goals rather than optimise many objectives as the basis for distinction between GP and the multiobjective programming approach, as explained in the next chapter.

Suggestions for further reading

Although for most researchers the starting point for GP is Charnes and Cooper (1961, App. B), the same authors have pointed out elsewhere (Charnes and Cooper 1975) that GP was actually originated in the 1950s to obtain 'constrained regression' estimates for an executive compensation plan (Charnes, Cooper and Ferguson 1955). These contributions can be regarded as having given birth to the fundamentals of the GP approach. Since then, however, an extensive body of literature has emerged on the development and application of the GP approach.

Specialist books devoted entirely to GP were written by Ijiri (1965), Lee (1972), Ignizio (1976, 1985), Schniederjans (1984), Romero (1991) and Schniederjans (1995).

Three bibliographical surveys of GP can also be cited: Lin (1980), Zanakis and Gupta (1985) and Romero (1986). Several state-of-the-art overviews of GP are worth recommending: Kornbluth (1973), Charnes and Cooper (1977), Nijkamp and Spronk (1978) and Ignizio (1978, 1983).

From the point view of using an efficient and reliable algorithm, the recently developed intelligent linear and integer GP system known as GPSYS (Jones, Tamiz and Mirrazavi 1998) should be cited. This software comes with analytical tools such as efficiency detection and restoration, redundancy checking, normalisation and interactive facilities.

For critical assessments of possible weaknesses of GP see Zeleny (1981) with a certain bias against GP (mainly against the LGP variant) and Hannan (1984, 1985) for a more balanced review, and Tamiz, Jones and Romero (1998) for an updated state-of-the-art survey. A critical analysis of the possibilities of GP in agriculture planning can be seen in Romero and Rehman (1984), which has provided the skeleton around which this chapter was written. References on WGP applications in agricultural planning include Wheeler and Russell (1977), Bazaraa and Bouzaher (1981) and Barnett et al. (1982). References of LGP applications in agricultural planning are: Barlett and Clawson (1978), Marten and Sancholuz (1982) and Dobbins and Mapp (1983). For critical exchange of views about the pros and cons of GP in agricultural planning see Romero and Rehman (1983) and McCarl and Blake (1983), Drynan (1985) and Romero and Rehman (1985b). Finally, Rehman and Romero (1993) places GP within the MCDM paradigm with reference to its applications to decision-making in agricultural and natural resource management.

Notes

1 We must warn here against the possible confusion regarding the apparent similarity between the slack and disposal variables in an LP model and the deviational variables as employed in GP. Admittedly there is a strong mathematical resemblance between the deviational and slack/disposal variables, but the functions they perform and meanings attached to them are quite different. The disposal/slack variables are essentially mathematical devices used to reduce the inequalities to equations to make the simplex algorithm work. The deviational variables on the other hand measure the under- or over-achievement of a particular goal for which they have been introduced into the GP model. Even if, in some instances, the numerical values for slack/disposal and deviational variables happen to coincide, their respective roles remain different and distinct.

2 The existence of alternative optima can be established easily from the final simplex tableau. If in that tableau there is at least one non-basic variable with a zero reduced cost, then alternative optimum solutions exist.

3 It is interesting to point out that if the NPV of the first investment is £6,700/ha, problem 3 will present alternative optimum solutions (all the points of the segment BC of Figure 3.2) and it will be necessary to formulate problem 4. However, after solving it we will obtain the point B as the optimal solution again.

4 Indeed an LP formulation is an LGP model where all the goals except one are included in the first priority as absolute ones in the sense that they must be satisfied to produce a feasible solution. The remaining singular goal (or the objective function to be optimised) is included in the second priority, while the goals that make up the first priority are the constraints of the conventional LP.

Chapter four
Multiobjective programming

It is quite a common decision-making situation where definite goals for the achievement of multiple objectives are not known in sufficient detail for them to be expressed as targets. In such situations, the general MCDM framework that has proved useful is multiobjective programming, known by the acronym MOP. This chapter introduces and explains MOP or vector optimisation techniques. The main purpose of MOP methods is to establish the set of efficient or Pareto optimal solutions in the sense explained in Chapter 2. The Pareto optimal feasible solutions are distinguished from the non Pareto-optimal ones and the traditional concept of optimum is replaced by the idea of efficiency or non-dominancy.

The main aspects and concepts of MOP are explained before analysing in detail the most widely used techniques to generate the efficient set, taking into account the computational problems associated with these methods. The problem of finding the optimum (compromise) solution from the efficient set, which is a corollary to the MOP approach, is examined in the next chapter.

An approximation of the multiobjective programming problem

The MOP or vector optimisation techniques help solve the problem of simultaneous optimisation of several objectives subject to a set of constraints, which are usually linear; it seeks to identify the set that contains efficient (non-dominated or Pareto optimal) solutions, as an optimum solution for several simultaneous objectives is not defined. The elements of this efficient set are the feasible solutions with the property that there are no other feasible solutions that can achieve the same or better performance for all the objectives, and strictly better for at least one objective, as explained in Chapter 2. As the purpose of MOP is to generate the efficient set, the general nature of problem can be stated as:

$$\underline{Eff} Z(\underline{x}) = [Z_1(\underline{x}), Z_2(\underline{x}), ..., Z_q(\underline{x})] \qquad (4.1)$$

subject to $\underline{x} \in \underline{F}$ where $\underline{Eff}$ means the search for the efficient solutions (in a minimising or a maximising sense) and $\underline{F}$ represents the feasible set.

To illustrate the use of MOP we revisit the example from Chapter 3. Supposing now that the farmer has two objectives, Z_1 and Z_2, that is: (1) to maximise the net present value of investment in plantation, and (2) to minimise the number of hours of casual labour hired for harvesting. The restraints of the problem are given by the set of equations given by the expression (3.1), augmented with a constraint specifying a minimum plantation area of 10 ha. The structure of this multiobjective model is given by (4.2) below where the signs for the coefficients of labour-hiring activities in Z_2 have been reversed in order to establish the efficiency of both the objectives in a maximisation sense.

$$\textit{Eff}\, Z(\underline{x}) = [Z_1\,(\underline{x}), Z_2\,(\underline{x})] \tag{4.2}$$

where

$$Z_1\,(\underline{x}) = 6250x_1 + 5000x_2$$
$$Z_2\,(\underline{x}) = -400x_1 - 450x_2$$

subject to

$$
\begin{aligned}
500x_1 + 400x_2 &\leq 15{,}000 \\
570x_1 + 575x_2 &\leq 22{,}000 \\
1{,}050x_1 + 825x_2 &\leq 29{,}000 \\
1{,}375x_1 + 1{,}025x_2 &\leq 36{,}000 \\
120x_1 + 180x_2 &\leq 4{,}000 \\
35x_1 + 35x_2 &\leq 1{,}000 \\
x_1 + x_2 &\geq 10 \\
\underline{x} &\geq \underline{0}
\end{aligned}
$$

As pointed out in the previous chapter, the fourth restraint embodies the first three, that is, they are redundant and are therefore ignored in what follows.

In this simple example, involving only two decision variables and two objective functions, it is possible to solve and to interpret the MOP problem graphically. Thus, the feasible set F can be represented by the polygon ABCDE in Figure 4.1. The five extreme points of this region along with the values for both objectives are shown in Table 4.1.

Table 4.1 Extreme points of the feasible set F

Extreme points	Decision variables		Objective functions	
	Pear trees (ha) x_1	Peach trees (ha) x_2	Z_1 (NPV) (£)	Z_1 (Casual labour) (hours)
A	10	0	62,500	4,000
B	26.18	0	163,625	10,472
C	19.18	9.38	166,775	11,893
D	0	22.22	111,111	10,000
E	0	10	50,000	4,500

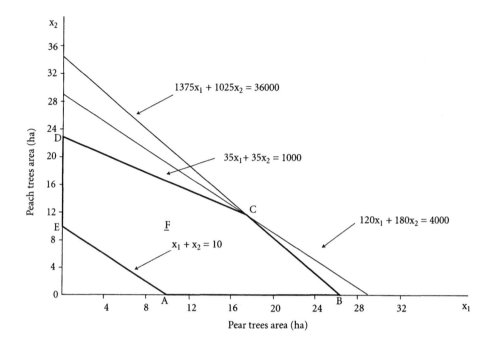

Figure 4.1 The feasible set in the decision variables space

The values achieved for the objectives Z_1 and Z_2 at the extreme points in the decision variables space generate new extreme points in the objectives space. The five extreme points of our example in the objective space are plotted in Figure 4.2. It should be noted that the points in the objective space are algebraic images of the points in the decision variable space. Thus, point A of Figure 4.1 maps point A' of Figure 4.2 through the values of Z_1 and Z_2 generated by point A. Points A', B', C', D' and E' of Figure 4.2 are connected by straight lines that are algebraic images of the straight lines connecting points A, B, C, D and E of Figure 4.1. In short, the feasible domain $\underline{F}'$ in the objectives space is a transformation (mapping) of the domain $\underline{F}$ in the decision variables space.

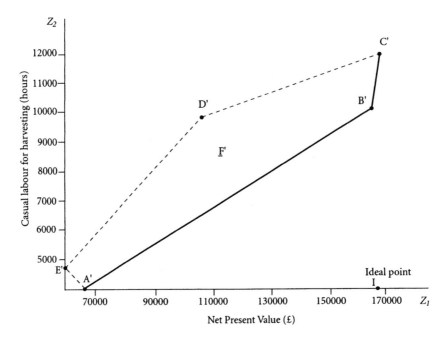

Figure 4.2 The image of the feasible set in the objectives space

On examining Figure 4.2, it is easy to deduce that the segments connected, A', B' and C', represent the efficient set in the objective space for the problem being analysed. In fact, the points of F that do not lie on the boundary formed by A', B' and C' are inferior or non-efficient because they offer less NPV and equal (or more) casual labour for harvesting or equal (or less) NPV and more casual labour for harvesting than any point belonging to the boundary itself. In the above example, therefore, the efficient set is given by the boundary ABC in the decision variables space or by the boundary A'B'C' in the objectives space.

The efficient set can be interpreted as the transformation curve, which measures the relationship between the two attributes. The slopes of the two segments A'B' and B'C'represent the trade-offs (or opportunity costs) between the two attributes; thus, the trade-off between NPV and casual labour along the portion A'B', according to the definition given in Chapter 2, equals:

$$T_{A'B'} = \frac{163,625 - 62,500}{10,472 - 4,000} = 25.28 \text{ £/hour}$$

The above trade-off indicates that over the segment A'B' each hour of hiring casual labour generates £25.28, or the opportunity cost of one hour of casual labour is worth £25.28 of

NPV. If this trade-off is worthwhile for the DM the point B' is preferred to the point A' and, if it is not then the point A' is preferred to the point B'.

When the number of decision variables is larger than two, this kind of graphical analysis cannot be used. In these situations multiobjective techniques are necessary to generate or at least to approximate the efficient set. There are basically three approaches to undertake that task: (1) the constraint method, (2) the weighting method and (3) the multiobjective simplex method. Of these three approaches only the last one can obtain an exact representation of the efficient set. However, as the applicability of the multiobjective simplex is limited to very small-sized MOP problems, its practical usefulness is restricted; therefore, it will not be presented here. In the next two sections the constraint and the weighting method are explained in detail, but before that the crucial intermediate concept of the pay-off matrix is described.

The pay-off matrix in MOP

One way of obtaining the initial and useful information of an MOP problem is to optimise each of the objectives separately over the efficient set and then to compute the value of each objective for each of the optimal solutions. This procedure generates a square matrix, called the 'pay-off matrix', as shown in Table 4.2 for the two objectives of our problem. The elements of the first row mean that the maximum NPV (£166,755) corresponds to hiring of 11,893 hours of casual labour. The elements of the second row mean that the minimum causal labour (4,000 hours) corresponds to an NPV of £62,500.

Table 4.2 Pay-off matrix for the two objectives

	Net present value (£)	Casual labour (hours)
Net present value	166,755	11,893
Casual labour	62,500	4,000

The degree of conflict between the two objectives can be investigated from the above pay-off matrix; thus in our example there is a clear conflict between the objectives NPV and casual labour. In fact, the maximum NPV is compatible only with a level of casual labour almost three times higher than its minimum level, and the minimum casual labour is compatible only with an NPV less than half of its maximum level.

In the literature on MOP, the elements of the main diagonal in the pay-off matrix are referred to as the 'ideal point'; that is, the solution where all the objectives achieve their optimum value. In our example the 'ideal point' contains £166,775 of NPV and 4,000 hours of casual labour. When the objectives are in conflict, as is the case in our example, the 'ideal point' is infeasible. This vector however is an essential reference point in the compromise programming approach as explained in the next chapter.

When we take the worst element – the maximum element of the objective is minimised, or the minimum element of the objective is maximised – from each row of the pay-off matrix, then we have what is called the 'anti-ideal' or 'nadir point'. This is the situation where all the objectives achieve their worst values. In our example the 'anti-ideal' point is £62,500 of NPV and 11,893 hours of casual labour. Although the 'anti-ideal' point is non-efficient, it has an important use in normalising objective functions measured in different units and with different absolute values, as described in the next chapter. Moreover, the difference between the ideal and the anti-ideal values define a range of values for each objective function, which is useful to operationalise the constraint method, which is explained below.

The constraint method

The basic idea here is to optimise one of the objectives while the others are specified as restraints. The 'efficient set' is then generated by parametric variation of the right-hand side elements of those restraints that have been built to represent the objectives. This method of generating the 'efficient set' was first introduced by Marglin (1967, pp. 24–25). So, for an MOP problem with q objectives to be maximised, the constraint method leads to the following:

$$\text{Maximise } Z_k (\underline{x}) \tag{4.3}$$

subject to

$$\underline{x} \in \underline{F}$$
$$Z_j (\underline{x}) \leq L_j \qquad j = 1, 2, \ldots, k - 1, k + 1, \ldots, q$$

where $Z_k(\underline{x})$ is the objective to be optimised. Through parametric variations of the right-hand sides L_j the efficient set is generated. Thus, for our example if the NPV is chosen as the objective to be optimised the application of the constraint method leads to the following parametric LP:

$$\text{Maximise } 6{,}250x_1 + 5{,}000x_2$$

subject to

$$\underline{x} \in \underline{F} \text{ [technical constraints from the model (4.2)]} \tag{4.4}$$
$$400x_1 + 450x_2 \leq L_1$$

The ideal and anti-ideal values for the casual labour objective shown in the second row of the pay-off matrix can be considered the upper and lower bounds for the range over which the parameter L_1 can vary; that is, between 4,000 and 11,893 hours/ha of casual labour.

By parameterising L_1 for values belonging to the interval [4,000, 11,893] an approximation of the efficient set is obtained. Table 4.3 shows the efficient points for our example. They are the extreme efficient points A, B and C in the decision variables space, or A', B' and C' in the objective space, and the interior points connecting these extreme efficient points.

Table 4.3 Extreme efficient points generated by the constraint method

Pear trees (ha) x1	Peach trees (ha) x2	Z1 (NPV) (£)	Z2 (Casual labour) (hours)	Right-hand side parameter (L1)
19.18	9.38	166,775	11,893	11,893
23.59	3.47	164,788	11,000	11,000
26.05	0	163,713	10,500	10,500
26.18	0	163,625	10,472	10,472
25	0	156,250	10,000	10,000
22.50	0	140,625	9,000	9,000
20	0	125,000	8,000	8,000
17.5	0	109,375	7,000	7,000
15	0	93,750	6,000	6,000
12.50	0	78,125	5,000	5,000
11.25	0	70,312	4,500	4,500
10	0	62,500	4,000	4,000

It should be noted that the constraint method guarantees efficient solutions only when the parametric constraints are binding in the optimal solutions, as is the case in our example. On the contrary, if for some values of the parameter L_k in the optimum solution, any of the parametric constraints are not binding and there are alternative optima, then the optimal solution provided by the constraint method may be inferior or non-efficient (Cohon 1978, pp. 117–118).

The constraint method requires p^{q-1} computer runs of the corresponding LP model, q being the number of objectives and p the number of sub-intervals over which the domain or the range of the objectives treated as restraints have been divided. Thus, a MOP problem with four objectives, giving five values to each of the objectives treated as restraint, will require $5^{4-1} = 125$ computer runs. This demanding computing task can, in practice, be undertaken by using parametric LP codes.

The weighting method

The basic idea here is to combine all the objectives into a single objective function. Each objective function is given a weight before all the objectives are added. Subsequently, the efficient set is generated through parametric variation of weights, as first introduced by Zadeh (1963). Thus, for a MOP problem with q objectives to be maximised, it leads to the following mathematical programming structure:

$$\text{Maximise } w_1 z_1\,(\underline{x}) + w_2 z_2\,(\underline{x}) + \ldots + w_q z_q\,(\underline{x})$$

subject to   (4.5)

$$\underline{x} \in \underline{F}$$
$$\underline{w} \geq \underline{0}$$

Through parametric variations of the weights w the efficient set can be generated. It should be noted that the weighting method guarantees efficient solutions only when the weights are larger than zero ($\underline{w} > \underline{0}$). If one of the weights is zero, and there are alternative optimal solutions, then the corresponding optimal solution provided by the weighting method can be inferior or non-efficient (Cohon 1978, pp. 109–110). Further, the weighting method can only generate extreme efficient points and not both the extreme and interior ones as the constraint method does.

For our example the application of the weighting method leads to the following parametric LP model:

$$\text{Maximise } w_1\,(6{,}250 x_1 + 5{,}000 x_2) + w_2\,(-400 x_1 - 450 x_2)$$

subject to   (4.6)

$$\underline{x} \in \underline{F}\,[\text{technical constraints from the model (4.2)}]$$
$$\underline{w} \geq \underline{0} \text{ and } w_2 \geq 0$$

Working with normalised weights (that is, making $w_1 + w_2 = 1$) and parameterising their values, the following results were obtained. For, $0.4 \leq w_1 \leq 1$ and therefore for $0 \leq w_2 \leq 0.6$ the optimum solution corresponds to point C (in the decision variables space) or C' (in the objectives space). For, $0.1 \leq w_1 \leq 0.4$ and therefore $0.6 \leq w_2 \leq 0.9$, the efficient solution changes to the points B or B' according to the space of reference considered. Finally, for $0 \leq w_1 \leq 0.1$ and therefore $0.9 \leq w_2 \leq 1$, the efficient solution changes again now to the points A or A' according to the space of reference.

It is tempting to interpret these weights as measures of the relative importance (or of preference) attached to each objective by the DM. Thus for instance, the solution C-C' that was obtained with the weights, $0.4 \leq w_1 \leq 1$ and $0 \leq w_2 \leq 0.6$, could be interpreted as the best solution for the DM who gives an importance of at least 0.66 times to the NPV objective as compared to the objective of hiring casual labour. That interpretation of the weights is correct only if the utility function for the DM is linear and additive; that is, it corresponds to the objective function stated for the MOP model in the expression (4.5) above. Obviously, that assumption is neither satisfied easily nor is valid generally.

Generally speaking these weights, ws, cannot be interpreted in a general way as measures of the relative importance given to each of the objectives. They must be treated only as parameters that can be varied systematically to generate the efficient set. It should also be pointed out that MOP does not get involved in considering the preferences of a DM; it can only partition the feasible set into efficient and non-efficient feasible solutions. However, in the second stage in the application of the MCDM approach, as presented in the next two chapters, the preferences of the DM are included in the analysis in order to choose the optimum or the best-compromise point from the efficient feasible solutions.

In common with the constraint method, the weighting method also requires p^{q-1} computer runs of the corresponding LP models, p now being the number of values given to the weights; again the number of the computer runs is an exponential function of the number of objectives. However, with the weighting method the coefficients of the objective function are parameterised instead of the right-hand sides values as in the constraint method. This makes the computations more difficult and therefore computationally the constraint method is a little more attractive than the weighting method.

The possibility that some efficient points may remain unexplored is common to both of the above generating techniques. To avoid that happening, it is necessary to reduce the scale of weights or the size of the sub-intervals of the right-hand sides considerably. A generating technique that has been derived from the weighting method, to avoid this drawback of the MOP models, is presented in the next section.

The noninferior set estimation method

Cohon et al. (1979) have proposed an MOP method called the noninferior set estimation (NISE) that permits a quick and good approximation of the efficient set when the number of objectives under consideration is usually no more than two. For problems of moderate size, the NISE method can generate the efficient set exactly. Although after certain modifications the NISE method could be applied to problems of higher dimensions, its real effectiveness diminishes considerably when more than two objectives are considered.

The operational aspects of the NISE method are now presented by using it to obtain the efficient set for the example from the previous section. The first step is to derive the pay-off matrix for the objectives under consideration, as done already in Table 4.2. The two rows of the pay-off matrix represent two important and singular points for the NISE method. In fact, as point A' is the efficient point with the best (lowest) value for the objective of casual labour minimisation and point C' is the efficient point with the best (highest) value for the objective of net present value maximisation, then segments A'G and C'G (see Figure 4.3) represent upper bounds of the efficient set. In other words, in the first step of the NISE method determines that the efficient set lies within or on the boundary of the triangle A'GC'.

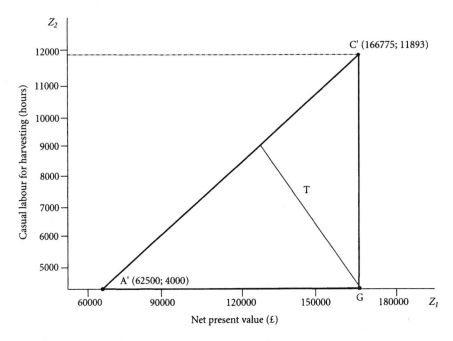

Figure 4.3 The application of the NISE method to the example problem: first iteration

The purpose of the NISE method is to reduce gradually the size of the zone of efficiency (i.e. triangle A'GC' for our example) through the iterative use of the weighting method. With that purpose the weights are chosen so that their quotient is equal to the slope of the segment connecting the extreme efficient points obtained in the previous iterations (points A' and C' for our example). In this way the next efficient point is the feasible solution farthest out in a direction perpendicular to the segment A'C'. Thus, in our case the weights w_1 and w_2 must satisfy:

$$\frac{w_1}{w_2} = \frac{11,893 - 4,000}{166,775 - 62,500} = 0.07569 \text{ (slope of the segment A'C')}$$

Now one of the weights, for instance w_2, can be set to an arbitrary value such as 1, w_2 then being 0.07569. Therefore, the objective function of the weighted problem to be considered now is given by:

Maximise $0.7569Z_1(\underline{x}) - Z_2$ (4.7)

We substitute the exact equation for Z_1 and Z_2 and manipulate them in (4.7) above, incorporating the corresponding constraints to obtain the following LP problem:

Maximise $73.06x_1 - 71.55x_2$

subject to (4.8)

$\underline{x} \in \underline{F}$ (technical constraints from the model (4.2)]

Solving the above LP problem, and substituting the optimum values obtained for x_1 and x_2 in $Z_1(\underline{x})$ and $Z_2(\underline{x})$, we have the point B' as the feasible solution farthest out in the direction perpendicular to A'C'. The information generated from this iteration is shown in Figure 4.4. Thus, the segments A'B' and C'B' represent the new lower bounds of the efficient set. In other words, we know now that the efficient set must lie within or on the boundaries of the triangles A'HB' and B'IC'.

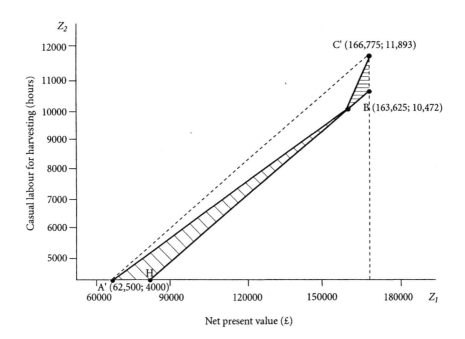

Figure 4.4 The application of the NISE method to the example problem: second iteration

The third iteration is another attempt to reduce the zone of efficiency again, where two new weighted LP problems are computed. In the first LP, the weights are calculated according to the slope of the segment C'B' and in the second one according to the slope of the segment A'B'. But in our example, essentially due to its simplicity, the solutions to these LP problems reproduce points C' and A' yet again; therefore, the iterative process can be stopped and the efficient set is given by the segments A'B' and B'C'.

For real problems, with a large number of decision variables and constraints, it would be tedious and rather impractical to try to obtain an exact representation of the efficient set. In such cases it should be enough to obtain an approximation of that set. To do that it should be sufficient to set *a priori* maximum allowable error as a percentage of the longest line perpendicular to the segment connecting the two efficient points which establish the first zone of efficiency; in our case, it is given by T; that is, the longest line perpendicular to the segment connecting the efficient points A' and C'. Therefore, the maximum allowable error can be expressed as a certain percentage of T. In this way, when in a given iteration the error obtained is less than the maximum allowable error then the process can be stopped without undertaking further iterations. Chapter 8 provides an illustration of the application of the NISE method, to obtain an exact representation of the efficient set for an agricultural planning problem with two objectives.

Multigoal programming

Before discussing some issues related to the use of the multiobjective programming (MOP) techniques we provide a brief outline of a related MCDM method, called multigoal programming (Zeleny 1982, pp. 298–300). Multigoal programming is really a hybrid between goal programming (GP) and MOP. It integrates the 'satisficing' idea from GP and the search for 'efficiency' from MOP; therefore, multigoal programming lies in the 'middle of the road' between MOP and GP and it operates by minimising the deviational variables as in vector or MOP optimisation but not lexicographically.

To illustrate the formulation of a multigoal programming problem we use our example by setting the levels for NPV and casual labour at £166,775 (i.e. the maximum NPV) and 6,000 hours respectively. Further, 1,000 hours of own tractors are considered as a target. In multigoal programming the efficiency with which all the objectives are achieved must be established in a minimising sense.

The multigoal programming formulation of our example then is:

$$\underline{\mathit{Eff}}\, Z(\underline{n}, \underline{p}) = [z_1(\underline{n}, \underline{p}), z_2(\underline{n}, \underline{p}), z_2(\underline{n}, \underline{p}), z_3(\underline{n}, \underline{p})]$$

where

$$z_1(\underline{n}, \underline{p}) = p_1$$
$$z_2(\underline{n}, \underline{p}) = p_2 \tag{4.9}$$
$$z_3(\underline{n}, \underline{p}) = p_3$$

subject to

$$1,375x_1 + 1,025x_2 \leq 36,000$$
$$x_1 + x_2 \geq 10$$
$$120x_1 + 180x_2 \leq 4,000$$
$$400x_1 + 450x_2 + n_1 - p_1 = 6,000$$
$$35x_1 + 35x_2 + n_2 - p_2 = 1,000$$
$$6,250x_1 + 5,000x_2 + n_3 - p_3 = 166,775$$
$$\underline{x} \geq \underline{0}, \underline{n} \geq \underline{0}, \underline{p} \geq \underline{0}$$

The real attraction of multigoal programming lies in combining a decision maker's desire to satisfy several goals via GP with the powerful and theoretically sound concept of efficiency in MOP; thus, avoiding some of the problems associated with GP formulations as pointed out in the last chapter. Two examples of the applications of multigoal programming to agricultural decision-making problems in agriculture are Lara and Romero (1992) in livestock diet formulation and Díaz-Balteiro and Romero (1998) in forest management.

Some issues related to the use of MOP techniques

Most of the weaknesses and drawbacks of the MOP approach are of operational and computational nature. In fact, the methods presented here, with the possible exception of NISE, can approximate the efficient set effectively only for a problem of moderate size. The only method that guarantees the exploration of all the extreme efficient points is the multiobjective simplex, which involves finding all the extreme efficient points by moving from one extreme efficient point to an adjacent efficient point. The algorithms for the multiobjective simplex that are available are mathematically sophisticated and require a large amount of computer time; the method has practical relevance only for small problems, thus limiting the possibilities of its application.

It should be pointed out that in most cases we are referring only to extreme efficient points or corner solutions, and they are of two types: (a) extreme points, (b) interior points (those which connect extreme efficient ones). The efficient set of solutions for the MOP model is derived from a convex and continuous region of the constraint set and, as such, it contains an infinite number of interior efficient points. But, from a practical point of view, only the subset of extreme efficient points is of interest. To find the whole efficient set is therefore not necessary for most real problems. There are, however, some algorithms that search for efficient faces; that is, the convex combination of extreme efficient points (e.g. Yu and Zeleny, 1975) and the whole efficient set is defined as the union of efficient faces.

Another operational problem is the huge number of extreme efficient points generated even for problems of moderate size. Several authors have reported applications where a few objectives and less than 50 or so variables and constraints have resulted in generating several hundred extreme efficient points. Of course, such a situation is undesirable for the decision maker (DM) as he is inundated with an undue amount of information making it almost impossible to make a choice.

Several approaches have been suggested to mitigate this problem. Steuer (1976) advocates the use of interval criterion weights rather than fixed ones as the weighting method. With this approach only that part of the efficient set, which is of greatest importance to the DM is analysed, so that a substantial amount of computer time is saved and also the size of the efficient set is reduced considerably.

Another method to reduce the size of the efficient set has been suggested by Steuer and Harris (1980) using filtering techniques. This kind of *pruning operation* discards efficient solutions that are not sufficiently different from other efficient solutions already calculated and retained by the filter. Once again the size of the efficient set can be reduced considerably. This is explained in some detail in Chapter 8 where in an application of MOP a set of fifty efficient points is generated and then reduced to a manageable cluster of only seven efficient points after filtering.

The MOP approach presented in this chapter can be regarded as the first stage in a decision-making process. In fact, using any of the techniques discussed above, the feasible solutions can be divided into two subsets: the efficient and the non-efficient ones. On this first stage the DM's preferences are not introduced at all. Once the inferior or non-efficient solutions are eliminated, there still remains for the DM the problem of choosing from the efficient solutions the optimum one. This task, the second stage of the decision-making process, can be undertaken in different ways, but always after having introduced the DM's preferences. In the next two chapters, two of the most widely used approaches to resolve this problem – compromise programming and interactive techniques – will be presented.

Suggestions for further reading

The basic ideas of MOP have been around for some time; Koopmans defined the concept of efficient or noninferior set as far back as in 1951. In the same year Kuhn and Tucker formulated the MOP problem and derived conditions for the existence of efficient solutions. These ideas have been gradually developing amongst researcher. A good account of these developments is given in Cochrane and Zeleny (1973), but since the appearance of this book the MOP approach has evolved considerably.

In the last few years many MCDM books have appeared with main emphasis on the MOP approach and the following ones can be recommended as easily readable material: Cohon (1978), Goicoechea et al. (1982) and Zeleny (1982). Other excellent books but requiring a higher level of mathematical training on the part of the reader include Zeleny (1974), Chankong and Haimes (1983), Yu (1985), Szidarovszky et al. (1986), Steuer (1986), Ringuest (1992), Miettinen (1999) and Ehrgott (2000).

Three bibliographical surveys that are devoted mainly to MOP are: Zeleny (1982, pp. 518–554), Stadler (1984, pp. 223–328) and Yu (1985, pp. 361–381). Similarly, the following state-of-the-art reviews on MOP can also be recommended: Roy (1971), Cohon and Marks

(1973), Roy and Vincke (1981), Evans (1984), Buchanan (1986) and Gal (1986). Ehrgott and Gandibleux (2002) is perhaps the most extensive and updated survey of theoretical and applied aspects of MOP.

The multiobjective simplex was first proposed by Philip (1972). Other versions include Evans and Steuer (1973) and Zeleny (1973). In practical terms the most powerful software available for the multiobjective simplex is the ADBASE (Steuer 1995); it can compute problems of only a moderate size (that is around fifty decision variables and constraints and three objective functions).

Methods for establishing the whole efficient set from the set of extreme efficient points are presented in Yu and Zeleny (1975), Iserman (1977) and Ecker et al. (1980). For a comparative analysis of NISE with the weighting and the constraints method see Balachandran and Gero (1984). An extension of the NISE method for three objectives is made in Balachandran and Gero (1985). A MOP analysis with fractional objectives can be seen in Kornbluth and Steuer (1981) and Kornbluth (1984). Stancu-Minasian (1997) is perhaps the most extensive survey of fractional programming.

The leading references on the application of MOP applications to agricultural planning are very limited and include only Hitchens et al. (1978), Thampapillai and Sinden (1979), Vedula and Rogers (1981), Apland et al. (1984) and Romero et al. (1987).

Chapter five

Compromise programming

The next multiple criteria decision-making (MCDM) technique to be explained is compromise programming (CP). To some extent it is a natural and a logical complement to multiobjective programming (MOP). MOP seeks to obtain the Pareto-efficient subset from the feasible solutions for a multi-objective problem, assuming that whatever the preferences of the decision-maker (DM), if his behaviour is rational, his choice will belong to the Pareto-efficient subset. Nonetheless to determine that optimum solution somehow it is necessary to introduce the DM's preferences. Compromise programming does that in a very realistic way, without having to rely on the questionable assumptions of the traditional utility theory. The basic idea in CP is to identify an ideal solution in the sense as explained in the last chapter. This ideal or utopian solution is only a point of reference for the DM. CP assumes, quite realistically, that any DM seeks a solution as close as possible to the ideal point, possibly the only assumption made by CP about human preferences. To achieve this closeness, a distance function is introduced into the analysis. The important point to emphasise here is that the concept of distance is not used in its geometric sense, but as a proxy measure for human preferences. The idea of a distance metric or a family of distance functions is essential for the CP technique to work. We therefore provide an intuitive explanation of the idea of a distance function as a prelude to the exposition of the operational and computational aspects of CP. The concluding part of this chapter compares the MCDM techniques presented so far, that is goal programming, multiobjective programming and compromise programming, to point out their pros and cons and to establish their mutual links and relationships.

An intuitive introduction to the concept of distance measures

If someone has to find the distance d between two points, $x^1 = (x_1^1, x_1^2)$ and $x^2 = (x_2^1, x_2^2)$, defined in a Cartesian plane, the answer according to the Pythagorean theorem is:

$$d = [(x_1^1 - x_1^2)^2 + (x_2^1, x_2^2)^2]^{1/2} \qquad (5.1)$$

Thus, for example, the distance d between (2,3) and (10,8) using this theorem is:

$$d = [(92 - 10)^2 + (3 - 8)^2]^{1/2} = 9.43$$

Figure 5.1 shows the Pythagorean or the Euclidian concept of distance diagrammatically.

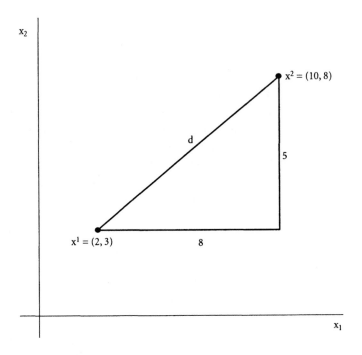

Figure 5.1 The Euclidean or Pythagorean distance between two points

This concept of distance can easily be extended to an n-dimensional space and the formula in (5.1) becomes:

$$d = [\sum_{j=1}^{n} (x_j^1 - x_j^2)^2]^{1/2} \tag{5.2}$$

Although this is the best-known measure of proximity between two points, it is not necessarily the only one. Thus, for some time now mathematicians using the notion of a family of L_p metrics or a family of distance measures have been able to provide a generalisation of the Euclidean distance as:

$$L_l = [\sum_{j=1}^{n} |x_j^1 - x_j^2|^{p}]^{1/p} \tag{5.3}$$

where the vertical lines represent absolute values. Obviously for each value of the parameter p, a particular distance is obtained. Thus, the Euclidean distance measure given above is a particular case of the family of L_p metrics when $p = 2$. For $p = 1$, expression (5.3) reduces to the following L_1 metric:

$$L_1 = \sum_{j=1}^{n} |x_j^1 - x_j^2| \qquad (5.4)$$

Thus, for our numerical example the L_1 distance will be:

$$L_1 = |2 - 10| + |3 - 8| = 13$$

Geometrically, the L_1 distance can be interpreted as the sum of the length of the sides of the triangle defined in Figure 5.1. When the parameter p has values greater than 2, it is not possible to give a geometrical interpretation to the distance measure; nonetheless, for such dimensions these distances can be computed. Thus, for instance, when a range of values is attached to p, the following distance measures are obtained for the two points of our example.

$$L_1 = |2 - 10| + |3 - 8| = 13$$
$$L_2 = |2 - 10|^2 + |3 - 8|^2 = 9.43$$
$$L_3 = |2 - 10|^3 + |3 - 8|^3 = 8.6$$
.
.
.
$$L_8 = |2 - 10|^8 + |3 - 8|^8 = 8.02$$
.
.
.
$$L_\infty = \text{Max} |2 - 10|, |3 - 8| = 8$$

where the Max signifies obtaining the largest absolute difference within the brackets. It is interesting to point out that as p increases more weight is given to the largest deviation. Thus, when $p = \infty$ the L_∞ distance is given exclusively by the largest deviation (i.e. $|2 - 10|$ = 8 for our example). In other words the parameter p weights the deviations according to their magnitudes.

From the data of our example it is easy to see that L_1 is the largest distance and L_∞ the shortest distance. This is a general property of the L_p metrics that is used extensively in developing the CP approach. We should however add that all these possible distance measures are bounded by the 'longest', the L_1 metric, and the 'shortest', the L_∞ metric, distances. This is also called the Chebyshev distance.

Obviously, in a strict two-dimensional geometric sense the use of L_p metrics for values of the parameter p greater than two is meaningless, because it would mean the existence of distances shorter than the straight line! However, the use of such metrics can be very useful, if they are used not in a geometric sense but as a measure for human preferences. In the following sections the L_p metrics are used to calculate 'distances' between solutions belonging to the efficient set and an ideal or a utopian point. The use of the distance concept as a proxy measure for human preferences makes CP approach a sound practical method in helping the DM to choose an optimum or the best-compromise solution from the efficient ones generated by MOP.

A discrete approximation of the best-compromise solution

The concept of ideal solution was introduced simultaneously by Yu (1973) and Zeleny (1973). Once this point has been defined, it is possible to establish the best-compromise solution as the nearest solution with respect to the ideal, accepting the basic postulate that the DM prefers solutions as close as possible to the ideal (*Zeleny's axiom of choice*). These simple but ingenious ideas form the basis of the CP approach.

Given the inherent conflict of multiple objectives, the ideal solution is infeasible; it is therefore necessary to look for compromise solutions. For that we need to calculate distances between each solution and the ideal point. The degree of closeness, d_j, between the jth objective and its ideal is given by:

$$d_j = Z_j^\star - Z_j(\underline{x})$$

when the jth objective is maximised, or as

$$d_j = Z_j(\underline{x}) - Z_j^\star$$

when the *j*th objective is minimised, $Z_j^\star$ being the ideal value for a given objective. The degrees of closeness between the different objectives and their ideal values are then added into a composite distance function. But as the units used to measure various objectives are different, therefore, in order to avoid a meaningless summation – adding pints of bitter to kilos of potatoes – the units of the measurement of the various objectives must be normalised. Not only that, if the absolute values for the achievement levels of the several objectives are different, then the scalarisation or normalisation of the degrees of closeness is necessary to avoid solutions biased towards those objectives that can achieve larger values. This problem, which is encountered in weighted goal programming (WGP) as well, can be overcome by using relative deviations rather than absolute ones. Thus, the degree of closeness d_j is given by:

$$d_j = \frac{|Z_j^* - Z_j(x)|}{|Z_j^* - Z_{*j}|}$$

where Z_{*j} is the anti-ideal or nadir point for the jth objective. The normalised degrees of closeness are bounded between 0 and 1; therefore, when an objective achieves its ideal solution then the degree of closeness is 0. On the contrary, when an objective achieves its anti-ideal solution then the degree of closeness is 1. Hence the normalised degrees of closeness measure the percentage of achievement of one objective with respect to its ideal value.

To measure the distance between each solution and the ideal point, CP uses the family of L_p metrics given by expression (5.5) as follows:

$$L_p(W) = \left[\sum_{j=1}^{n} W_j^p \left| \frac{Z_j^* - Z_j(x)}{Z_j - Z} \right|^p \right]^{1/2} \tag{5.5}$$

or what is equivalent to:

$$L_p(W) = \left[\left(\sum_{j=1}^{n} W_j d_j \right)^p \right]^{1/p} \tag{5.6}$$

where W_js are the weights representing the importance of the discrepancy between the jth objective and the ideal point; that is, W_j measures the relative importance of the jth objective in a given decision situation.

The family of distance functions (5.6) can be applied to a set of feasible and efficient alternatives in order to choose the best-compromise solution. Thus, the alternative with the lowest value for $L_p(W)$ will be the best-compromise solution because it is the nearest solution with respect to the ideal point. Obviously, the best-compromise solution can change according to the values of the parameter p and the weights W_j that are chosen by the DM. The parameter p acts as a weight attached to the deviations according to their magnitudes. Similarly W_js become the weights for various deviations signifying the relative importance of each objective. For different sets of values of p and W_j we can generate different compromise solutions.

We now apply this discrete version of CP to the agricultural planning example that has been introduced already in the previous chapters. Thus in Table 5.1 the distances between each extreme efficient point (A', B' and C') and the ideal point have been calculated for the three measures of distance L_1, L_2 and L_∞ and for several structures of weights. As an illustration the details of calculating the distance between point B? and its ideal, according to the L_2 metric for $W_1 = 3$ and $W_2 = 1$, are given below:

$$L_2(3,1) = \left[\ 3^2 \left| \frac{166{,}775 - 163{,}625}{166{,}775 - 62{,}500} \right|^2 + 1^2 \left| \frac{4{,}000 - 10{,}472}{4{,}000 - 11{,}893} \right|^2 \ \right]^{1/2} = 0.825$$

From Table 5.1 it can be seen that given the structure of weights W_j, of the three extreme efficient points B' is the one located nearest to the ideal point; whatever measure of distance is used. In other words, the point B' in the objective space or the point B in the decision variable space are the best-compromise solutions.

Table 5.1 Compromise programming (discrete approximation)

	A'	B'	C'	Z_j^*	Z_j^*
NPV (Z1)	62,500	163,65	166,775	166,775	62,500
Casual labour (Zj)	4,000	10,472	11,893	4,000	11,893
d_1	1	0.030	0		
d_2	0	0.820	1		
L_1 $W_1 = 1$	1	0.850	1		
L_2 $W_2 = 1$	1	0.820	1		
L_∞	1	0.820	1		
L_1 $W_1 = 2$	2	0.860	1		
L_2 $W_2 = 2$	2	0.860	1		
L_∞	2	0.820	1		
L_1 $W_1 = 3$	3	0.910	1		
L_2 $W_2 = 1$	3	0.825	1		
L_∞	3	0.820	1		

Compromise programming – a continuous setting

The discrete version of compromise programming (CP) is a useful device to rank a finite set of alternatives. It has, however, two possible weaknesses. First, such an application of CP requires that the efficient set (or at least the subset of extreme efficient points) has been determined already. Second, the best-compromise point should always be an extreme efficient point, which is not always the case as in many instances the best-compromises could be Pareto-interior points. Hence, in the discrete approximation a considerable amount of information is missed; therefore, in our example the interior points belonging to the efficient segment A'B' (see Figure 4.2) cannot be considered as compromises easily, particularly when some of these points could have been the best-compromise for a given metric and a set of weights indicating the preferences of the DM.

Both of these problems are avoidable if CP is used in a continuous setting. An added bonus is that the best-compromise solutions are obtained straightforwardly from conventional LP models. Thus, for the L_1 metric (i.e. $p = 1$) expression (5.5) gives the minimisation of the relative deviation with respect to the ideal. But, as $Z_j^* \geq Z_j(x)$ for every j, because Z_j^* is a component of the ideal vector, then the absolute value signs of (5.5) can be dropped and, therefore, for the L_1 metric the best-compromise or the closest solution to the ideal point can be obtained by solving the following LP problem:

$$\text{Min } L_1(W) = \sum_{j=1}^{n} W_j \frac{Z_j^* - Z_j(x)}{Z_j^* - Z_{*j}} \tag{5.7}$$

subject to

$$\underline{x} \in \underline{F}$$

where $\underline{F}$ is the feasible set. Applying the model (5.7) to the data of our example the best-compromise solution for the L_1 metric is obtained by solving the following LP problem:

$$\text{Min } L_1 = W_1 \frac{166{,}775 - Z_1(x)}{166{,}775 - 62{,}500} + W_2 \frac{Z_2(x) - 4{,}000}{11{,}893 - 4{,}000}$$

subject to

$$\underline{x} \in \underline{F} \text{ [technical constraints from model (2) of chapter 4]}$$

The optimum solution of the above LP problem for $W_1 = W_2$ (that is, when the objectives are equally important), is given by the point B'; therefore, the point B' is the best-compromise solution and this means that B' is the efficient point closest to the ideal point when the metric L_1 is used.

For the L_∞ metric ($p = \infty$), the maximum deviation from among the individual deviations is minimised. That is, when $p = \infty$ only the largest deviation counts. For this metric, the best-compromise solution is obtained by solving the following LP problem:

$$\text{Min } L_\infty = d$$

subject to

$$W_1 = \frac{Z_1^* - Z_1(x)}{Z_1^* - Z_{*1}} \leq d$$

$$\vdots \tag{5.8}$$

$$W_n = \frac{Z_n^* - Z_n(x)}{Z_n^* - Z_{*n}} \leq d$$

where d is the largest deviation[1]. On applying the model (5.8) to our example the best-compromise solution for the L metric is obtained from the following LP problem:

$$\text{Min } L_\infty = d$$

subject to

$$W_1 \frac{166{,}775 - Z_1(x)}{166{,}775 - 62{,}500} \leq d \tag{5.9}$$

$$W_2 \frac{Z_2(x) - 4{,}000}{11{,}893 - 4{,}000} \leq d$$

$x \in F$ [technical constraints from model (2) of chapter 4]

The optimum solution of the above LP problem, assuming once again that $W_1 = W_2$, is given by the point Z' of Figure 5.2. That point corresponds to a NPV of £119,000 and hiring 7,616 hours of casual labour. The image of Z' in the decision variable space has the coordinates $x_1 = 19.04$ and $x_2 = 0$; that is, the best-compromise for $p = \infty$ consists of planting 19.04 ha of pear trees and no peach trees.

Nonlinear algorithms are needed to obtain the best-compromise solutions for other metrics. However, Yu (1973, 1985 pp. 76–77) proved that for problems with two objectives L_1 and L_∞ metrics define a subset of the efficient set, which Zeleny (1974) calls the compromise set. The other best-compromise solutions fall between the solutions corresponding to L_1 and L_∞ metrics; therefore, the solutions provided by the two LP models formulated above characterise the bounds of the compromise set. Thus, segment Z'B' represents the compromise set for our example (see Figure 5.2).

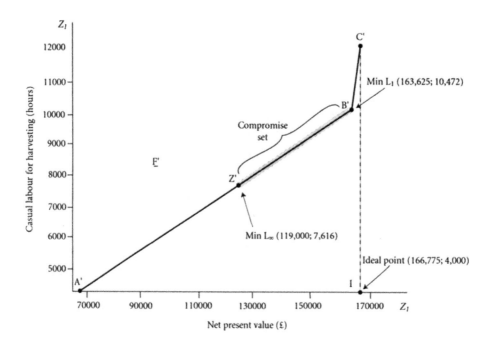

Figure 5.2 The ideal point and compromise set

For different sets of values for the weights W_1 and W_2 the structure of the compromise sets can be modified. A sensitivity analysis with the weights can furnish the DM with worthwhile information related to the stability of the solution and the range within which the compromise sets can be defined. This kind of sensitivity analysis is used in Chapter 7 where CP is used to analyse a risk programming problem in agriculture.

The method of the displaced ideal

According to the neoclassical marginal analysis, the optimum choice for the DM is given by the feasible solution where his utility function reaches a maximum value. In the context of our example that solution corresponds to the point of tangency between the efficient set (i.e. the boundary A'B'C') and the family of iso-utility curves defined in the NPV-casual labour space. But as already pointed out, establishing this family of iso-utility curves is a very difficult task in practice and is dependent on rather unrealistic assumptions.

In CP the underlying idea is that we cannot have a reliable mathematical representation of the DM's actual utility function in practice; therefore, CP does not even attempt to determine it and instead identifies a compromise set that can be interpreted as that portion of the efficient set where the tangency between the iso-utility functions and the efficient set will presumably occur; that is, the compromise set can be interpreted as something like a 'landing area' for the utility curve. For our example therefore any conceivable utility function defined in the NPV-casual labour space for the DM, given that the two attributes are equally important, would attain its tangency at one point of the segment Z'B', which defines the compromise set.

Taking the preceding fact into account, it is inconvenient to work with compromise sets that range over a big portion of the efficient set. In fact, with a large compromise set it is not easy for the DM to choose the optimum solutions from the efficient compromises; and also too large a neighbourhood of the tangency point provides little interesting information. It is therefore worthwhile to explore operational methods that allow us to reduce the size of the compromise sets. Zeleny (1974, 1976a) has suggested one such method called the displaced ideal which helps to reduce the compromise set to a manageable size. Let us now examine the basic ideas that underlie this method.

Assume that the compromise set Z'B' is considered to be too large; that is, to state that the tangency between utility function and the efficient set will presumably occur at some point of this segment may be regarded as a piece of information which is fuzzy and hardly valuable for the DM. The situation can however be redeemed if the size of the compromise set is reduced. The DM is asked to discard those portions of the efficient set that do not interest him at all. For instance, assume that the DM decides to discard efficient solutions belonging to the segment B'C' because they require too many hours of casual labour (more than 10,472 hours), and also the solutions along A'Z' as they provide too little NPV (less than £119,000).

Discarding these efficient points implies obtaining a new ideal point. In other words, the ideal point is displaced. Thus, in our example the discarding of that part of the efficient set gives the new ideal point I' (163,625; 7,616) represented in Figure 5.3. Obviously, the displacements of the ideal and the anti-ideal will produce a new compromise set. As it was explained in the preceding section, the bounds of the compromise set are established by solving two LP problems. For the L_1 metric the LP problem to be solved now is the following one:

$$\text{Min } L_1 = W_1 \frac{163,625 - Z_1(\underline{x})}{163,625 - 110,000} + W_2 \frac{Z_2(\underline{x}) - 7,616}{10,472 - 7,616} \tag{5.10}$$

subject to

$\underline{x} \in \underline{F}$ [technical mconstraints from model (2) of chapter 4]

Solving the above LP problem for $W_1 = W_2$, the L_1 bound B' of the compromise set is obtained again; that is, the displacement of the ideal has not produced any displacement in the L_1 bound of the compromise set. The L_∞ bound of the new compromise set is now obtained by solving the following problem:

Min $L_\infty = d$

subject to

$$W_1 \frac{163,625 - Z_1(x)}{163,625 - 119,000} \leq d$$

(5.11)

$$W_2 \frac{Z_2(x) - 7,616}{10,472 - 7,616} \leq d$$

$\underline{x} \in \underline{F}$ [technical constraints from model (2) of chapter 4]

Solving the above LP problem for $W_1 = W_2$, the new L_∞ bound of the compromise set is obtained. That bound is represented in Figure 5.3 as point Z". That point corresponds to a NPV value of £141,313 and hiring 9,044 hours of casual labour. The image of that point in the decision variable space specifies planting 22.61 ha of pear trees and no peach trees.

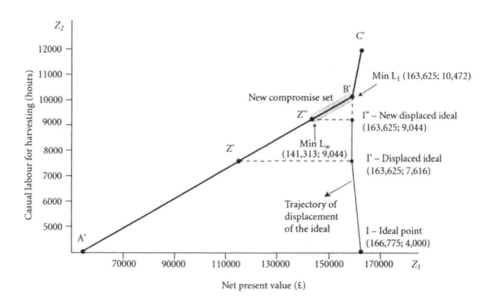

Figure 5.3 The trajectory of the displacement of the ideal and the new compromise set

Displacing the ideal has produced the segment Z"B, which is a smaller and much more manageable compromise set. It reduces the landing area for the utility function considerably and thus it is now much easier for the DM to choose a best-compromise solution from the new compromise set Z"B'.

Obviously, this method can be implemented iteratively. If the size of the compromise set obtained in the first iteration is not small enough, then the DM is asked to discard new portions of the efficient set, producing a new displacement of the ideal, continuing the iterative procedure until a satisfactory compromise set is obtained. Thus, if in our example the DM discards efficient solutions outside the new compromise set Z"B', the new ideal I" in Figure 5.3 is obtained, which will lead to a new contraction of the compromise set. This method is a forerunner to the interactive techniques discussed in the next chapter.

Pros and cons of GP, MOP and CP

Having discussed the nature and functioning of the various MCDM techniques, it is worth critically assessing the individual pros and cons of each method in the light of their different attributes.

The first attribute to consider is the amount of computational time required in using each technique. In this respect, GP is most efficient as it requires a single 'computer run'. If sensitivity analysis is carried out say for targets, weights, order of priorities, etc. then, of course, the computing time increases correspondingly. However, the MOP approach involves even more computation, as with both the weighting and constraint method algorithms, the number of 'computer runs' required is an exponential function of the number of objectives involved.

These demanding computing time requirements of MOP can be reduced significantly if CP (in a continuous setting) is used instead; as it is only necessary to solve two LP problems (for the metrics L_1 and L_∞) for each set of weights; thus giving us a part of the efficient set as the compromise set to work with. This however implies a possible weakness. We lose the information on the trade-offs between objectives that was contained in those parts of the efficient set that are excluded from the compromise set.

In terms of the quantity of information and the precision with which it is required from the DM, perhaps GP is the most difficult approach. The DM has to provide precise target values, weights attached to each deviational variable, pre-emptive ordering of preferences, etc. Much of this is very difficult to obtain in some cases, as with the target values, it could be argued that this is the kind of information that the model should provide to the DM rather than it being an input requirement.

The MOP approach is at the opposite end of the scale as regards the information needed from the DM. To build an MOP model, it is not necessary to know anything about the DM's preferences. The mathematical expression of the objectives being considered is sufficient. For CP we need to know only the relative preferences of the DM for each objective (that is, the weights attached to the discrepancies between each objective and their ideal values) in order to approximate the compromise set for the different metrics.

As regards the information produced by the model for use by the DM, GP is clearly inferior to the other two techniques. In fact, the different GP variants provide only a single solution; that is, the set of decision variables that permits the closest satisfaction of the different goals. Although a sensitivity analysis subsequent to obtaining the optimal solution may be sufficient in many instances, the fact remains that GP provides rather meagre information compared to either MOP or CP.

The efficient set generated by MOP furnishes us with extremely worthwhile information for the purposes of decision-making. This set contains the transformation curve or the trade-offs for the objectives involved – an essential piece of information for evaluating different alternatives before the DM makes a choice. This advantage of MOP is particularly useful when the number of objectives is limited to two or three, as it is then possible to display the efficient set and the corresponding trade-offs graphically; CP provides the same information as MOP by identifying the bounds of that portion of the efficient set which is closest to the ideal point. These considerations do not lead us to a definite conclusion about the superiority of one MCDM approach relative to others; perhaps that is not even necessary. As Ignizio (1983, p. 278) says, 'there is not now, and probably never shall be, one single "best" approach to all types of multiobjective mathematical programming problems'.

In agricultural planning involving multiple criteria decisions, the choice of a given MCDM approach as the modelling technique will inevitably depend upon several factors. If the decision-making problem being analysed involves many attributes, say six, and the constraint set is rather complex, then all the potential benefits of either MOP or CP vanish. A problem of this size is computationally intractable via MOP. If the efficient set is approximated via the weighting method and the number of array of weights used is five – a modest figure for six objectives! – then the number of 'computer runs' needed is the massive figure of $5^6 = 15,625$. Even if we manage to complete all these runs, still the information of efficient points, trade-offs, etc. that the DM is inundated with makes it almost impossible for him to digest and make sensible use of it. However, a decision-making problem of such dimensions can be easily accommodated within the GP framework.

On the other hand, if the problem being modelled involves no more than three attributes, and the DM is uncertain about the values of his targets and/or the specification of his preferences with respect to each attribute, then both MOP and CP would be more promising than GP. In any case a thorough analysis of the problem before the modelling process proceeds should determine the most appropriate technique to use.

Relationships between different MCDM approaches

In this section, by analysing some of the links and relationships that exist between the three MCDM approaches that have been introduced so far, certain theoretical conclusions are inferred, which have practical consequences.

First, a WGP model in which all the targets have been set as the ideal values of the goals (i.e. infeasible bounds), it is really a CP model with the L_1 metric without normalised objectives. Let us consider the following general WGP where the achievement of all the goals are assumed to be greater than or equal to their targets:

$$\text{Min} \sum_{j=1}^{q} W_j n_j$$

subject to

$$f_j(\underline{x}) + n_j - p_j = t_j \qquad\qquad j = 1, ..., q \qquad\qquad (5.12)$$

$$\underline{x} \in \underline{F}$$

where $\underline{F}$ as usual is the feasible set

In (5.12) if we set $t_j = t_j^*$ for every j, where t_j^* is an ideal value (maximum bound), then we have $p_j = 0$ because as t_j^* is a maximum bound, obviously the possible over-achievements are zero. Therefore, we have:

$$n_j = t_j^* - f(\underline{x}) \qquad\qquad j = 1, 2, ..., n \qquad\qquad (5.13)$$

The structure of (5.12) thus turns into the following model:

$$\text{Min} \sum_{j=1}^{q} W\left[t_j^* - f_j(\underline{x})\right] \qquad\qquad\qquad (5.14)$$

$$\underline{x} \in \underline{F}$$

The algebraic expression in (5.14) is evidently a CP model for the L_1 metric when the objectives are not normalised (cf. the expression (5.7)). It can, therefore, be said that in a way GP can be considered a model more general than CP, because GP can use as targets any vector of values when CP uses as targets the vector of ideal values. However, as CP uses any kind of metric whereas GP uses only the L_1 metric, it can be said that from this perspective CP is more general than GP.

In a CP model using the L_1 metric is tantamount to the weighting method of MOP with normalised objectives. In fact, it is quite simple to see that the minimisation of the objective function of (5.7) is equivalent to the maximisation of the following expression:

$$\text{Max} \sum_{j=1}^{n} W_j \frac{Z_j(\underline{x})}{Z_j^* - Z_{*j}} \tag{5.15}$$

This is the same as the objective function of the weighting method, when the objectives are normalised (see expression (4.5) in Chapter 4).

Finally, when L_∞ metric is used in CP, and the objectives are not normalised, the model is equivalent to a MINIMAX GP formulation (see Chapter 3) where the targets have been set as the ideal values of the goals. In fact, if in the structure (3.8) of Chapter 3 we set $t_j = t_j^*$ for every j, then $p_j = 0$, reducing it a simple algebraic manipulation to obtain the following model:

Min d

subject to

$$t_j^* - f_j(\underline{x}) < \text{d} \tag{5.16}$$

$$\underline{x} \in \underline{F}$$

Obviously, this structure in (5.16) is a CP model for the L_∞ metric when the objectives are not normalised. This theoretical analysis leads to some interesting consequences.

As has been pointed out in Chapter 4 that if the weights in the weighting method are interpreted as measures of the relative preference attached by the DM to each of the objectives, then the existence of an additive utility function is implied. So, because of the equivalence between CP and MOP the use of the L_1 metric in CP implies the use of that kind of restrictive utility function. Because of that, the concept of compromise set is especially useful in overcoming that weakness because that set is compatible with a wide range of forms of utility functions for the DM. Moreover, because of the equivalence between CP and WGP, this variant of GP also subsumes an additive utility function for the DM.

In the MCDM literature the models where the targets have been set as infeasible bounds (e.g. ideal values) are easily found. In those instances modellers are not actually utilising GP but the MOP technique without being aware of the situation. Thus the sensitivity analysis of the weights attached to the deviational variables is, in fact, an indirect way of approximating the efficient set. In this context, a WGP model provides the same solution as does a CP model for the L_1 metric where the objectives are not normalised. In the same way a MINMAX GP model has the same solution as a CP model for the L_∞ metric without normalising the objectives. Therefore, the rather common practice of setting all the targets in a GP model as the ideal values of the goals, despite having the practical advantage of securing efficient or non-dominated solutions, is not really GP but an unrecognised form of CP, implicitly assuming in some cases (as in the WGP variant) the existence of a very restrictive form of the DM's utility function.

Suggestions for further reading

The notion of an ideal comes from psychology and was introduced by Coombs (1958) as a point where the different attributes desired by a DM achieve an ideal value without the imposition of constraints on his choice. The first operational use of such an ideal solution is credited to Benayoun et al. (1971) for developing the interactive MOP technique known as STEM. CP was first introduced by Zeleny in 1973 and was subsequently refined by him (see Zeleny 1974, 1976a, 1982). He was also responsible for proposing the method of the displaced ideal (see Zeleny 1974, 1976a). A very readable account of this can be found in Zeleny (1982).

Some interesting further developments and extensions of CP have taken place. Yu (1973) has extended the concept of compromise solution to group level decision- making, while mathematical and computational developments in CP include the works by Yu and Leitman (1974), Yu (1985, chap. 4) and Gearhart (1979, 1984). Yilmaz (1984) has used the method of the displaced ideal in an uncertain environment. Michalowski (1981) has proposed a contraction approach to selection of the final compromise set without involving the DM. Freimer and Yu (1976) have shown that for problems with more than two criteria, points L_1 and L_∞ do not necessarily define a compromise set. Blasco et al. (1999) demonstrated that the boundedness of the compromise set by metrics $p = 1$ and $p = \infty$ for more than two criteria is guaranteed under very general conditions.

A pioneering application of CP in agricultural planning is that of Romero et al. (1987). In water resources planning the pioneer CP applications are Duckstein and Oprovic (1980) and Gershon and Duckstein (1983). and in interregional planning Hafkamp and Nijkamp (1983). A very recent survey by Hayashi (2000) provides updated information about CP applications in agriculture.

Comparative reviews of GP, MOP and CP can be found in Cohon and Marks (1975), Willis and Perlack (1980), Rehman and Romero (1987a) and Rehman and Romero (1993). Further elaboration of the mutual links and relationships involving the different MCDM approaches can be found in Romero (2001).

Note

1   As mentioned in Chapter 3, the MINMAX GP and CP are close relatives. Later in this chapter we show that model (5.8) is tantamount to a MINMAX GP, where all the targets are maximum bounds (e.g. the components of the ideal vector).

Chapter six
The interactive multiple criteria decision-making approach

The interactive multiple criteria decision-making (MCDM) approach as presented in this chapter implies a progressive evolution and definition of a decision maker's (DM) preferences through an *interaction* between him and the results generated from various runs of the model. This *interaction* becomes a dialogue in which the model responds to an initial set of the DM's preferences or trade-offs, and then when this response has been examined another set is offered, and thus the procedure progresses in an interactive and iterative way until the DM has found a solution that he regards as satisfactory. The interactive approach can be used for decision-making problems that involve analysing either multiple objectives or multiple goals. In the first case, the interaction attempts to find the optimum (the best compromise) solution when the DM provides his specific preferences (rather than the absolutes ones) with respect to the efficient solutions presented to him. Similarly, in the second case the values of some important parameters, such as targets, weights, and order of priorities, in the goal programming (GP) model are elicited from the DM through an *interaction* with the model involving successive computer runs.

The interactive methods have been one of the most productive areas of research in the MCDM field for some time. Despite that, the approach itself is not free of difficulties and drawbacks; among these, perhaps the main one is the mathematical sophistication that creeps in when some of these models are operationalised. This approach does however deserve to be explored in some detail as it does not require too many restrictive assumptions about the preferences of the DM, as several of the proposed interactive MCDM methods do not attempt to establish a strict mathematical representation of the DM's utility function. The information required is more modest as the general purpose of this approach is to obtain a local approximation relevant to a specific situation of the DM's utility function or the point of maximum utility through an interaction between him and the model, as shown in Figure 6.1.

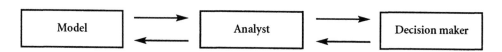

Figure 6.1 The interactive decision-making process

In this chapter, the general structure of an interactive MCDM process is presented first before examining three of the most promising methods in detail. An interactive MCDM approach should assist the DM in finding a compromise between the different attributes of a decision-making situation. It is this consideration that has determined the choice of the methods being presented here along with mathematical simplicity of the method, relevance to the agricultural planning problems and the amount of information required from the DM. We believe that acquiring an understanding of these three methods should be sufficient for the reader to appreciate the possibilities offered by interactive MCDM techniques.

Structure of an interactive MCDM process

Generally, an interactive process involves three entities or components: the DM, the analyst and the operationalised model. In this context, the analyst is a '*messenger*' or an '*intermediary*' between the model and the DM. The process of interaction, whatever the approach, can be summarised as follows.

The analyst obtains an initial solution from the model for presentation to the DM, who evaluates the solution in the light of his preferences. The DM is not required to give any information about his absolute preferences, such as the relative final weights attached to each objective, but only his current preferences with respect to the solution being presented to him by the analyst; such preferences that have been provided by the DM are fed back to the model by the analyst. A new solution is generated and the DM gives his revised current or local preferences, considering the new solution. This process continues in an iterative and interactive way, until a solution that is considered good enough by the DM has been obtained.

In short, all the interactive methods for multiobjective programming (MOP) or goal programming (GP) operate in an iterative way by moving from one efficient solution to another according to the direction determined by the preferences of the DM. We should emphasise that this approach does not make any general assumption about the DM's preference structure, for example, having to assume that maximizing net present value (NPV) is twice as important as minimising the use of hired casual labour. Instead the DM's preferences are extracted or elicited interactively via the *dialogue* with the computerised model.

All the interactive methods can be classified according to how the communication between the DM and the model takes place via the analyst; that is, according to the kind of information expected to be provided iteratively, by the DM during the interactive process. There are methods that require that the DM to provide only the information on the values of his local trade-offs between the objectives or goals under consideration. For instance, using the example from the preceding chapters, the analyst could have the following '*conversation*' with the DM:

Question. What is your trade-off between the objectives of NPV and casual labour?

Answer. For each hour of decrease in hiring casual labour I accept a decrease of £20 in NPV, therefore the trade-off between these two objectives is 20.

The model is then rerun with this information to obtain a new solution. The methods based on this type of communication between the participants are difficult to implement, for two reasons. In general, when a DM is asked to provide some quantitative idea of the trade-offs between his objectives he finds it difficult to commit himself to a numerical value. Not only that; the algorithms used by analyst in this situation are mathematically complex. We are therefore not presenting such methods here but the interested reader can consult Geoffrion et al. (1972).

In the second type of interactive method the DM is asked by the analyst if he accepts or rejects a certain set of trade-offs. In our example it would progress as follows:

Question. Do you accept an increase of £10,000 in NPV, along with an increase of 500 hours of casual labour and 100 hours of hired tractors?

Answer. Yes | No | I do not know.

The method proposed by Zionts and Wallenius (1976) as presented later corresponds to this scheme of communication.

Finally, there is a third type of interactive method where the DM is not asked directly or indirectly about his current trade-offs but instead his opinion on the acceptance or rejection of a given feasible efficient solution is solicited. If a solution is not acceptable, then the DM should indicate for which objectives the performance should be improved. Going back to our example, the following can be envisaged:

Question. Do you accept an investment plan providing a NPV of £150,000, 10,000 hours of casual labour and 300 hours of hired tractors?

If the DM's response is negative, then he must indicate which objective will have to be improved. For instance, he may regard that the values for casual labour and hired tractors are too high and should therefore be reduced, while the NPV could be degraded to £30,000 in order to improve the performance of the other two attributes.

The STEM method and the interactive multiple goal programming as discussed in later sections conforms to this scheme of communication. To a certain extent, the method of the displaced ideal of the preceding chapter can also be regarded as a similar interactive method. In explaining the three interactive MCDM presented here, we revisit the decision-making models used in the previous chapters, making modifications where necessary.

The STEM method

Benayoun et al. (1971) have proposed an interactive MCDM approach called STEM which is perhaps the oldest such method and is also one of the most widely used. We now explain the general structure of this method giving some idea of its algorithmic function before illustrating its application.

The STEM method proceeds in two phases: a calculation phase and a decision phase. The interaction between the DM and the model takes place in the second phase only in the form of the third scheme of communication as explained already.

The first step in the calculation phase is to generate the pay-off matrix in order to obtain the ideal, Z_j^*, and the anti-ideal, Z_{*j}, values for each of the objectives included in the model. Then a first approximation of the solution nearest to the ideal point, in a minimax sense (according to the L_∞ metric), is obtained (see the preceding chapter) by solving the following linear programme:

Min d

subject to

$$W_j[Z_j^* - Z_j(\underline{x})] \le d \qquad j = 1, 2, ..., q \tag{6.1}$$

$$\underline{x} \in F$$

The normalising weights W_js in the STEM method are defined as:

$$W_j = \frac{v_j}{\displaystyle\sum_{j=1}^{q} v_j} \tag{6.2}$$

where

$$v_j = \left[\frac{Z_j^* - Z_{*j}}{Z_j^*}\right] \cdot \left[\frac{1}{\sqrt{\displaystyle\sum_{k=1}^{m} c_{kj}^2}}\right] \tag{6.3}$$

if the jth objective is maximised

or

$$v_j = \left[\frac{Z_{*j} - Z_j^*}{Z_{*j}}\right] \cdot \left[\frac{1}{\sqrt{\displaystyle\sum_{k=1}^{m} c_{kj}^2}}\right]$$

If the jth objective is minimised.

In the above the expression c_{kj} represents the coefficients of the jth objective. This particular normalising process has some features that are worth noting. First, the weights W_js do not represent the DM's actual preferences; they are just the normalised weights. Second, the expression (6.2) ensures that the W_js sum to one, which is a very practical requirement for comparing different strategies. Third, the first terms of the equation (6.3) place relatively more weight on the objectives that deviate a great deal from the optimum solution (that is the difference between their ideal and anti-ideal values is large). Finally, the purpose of the second terms of (6.3) is to normalise the objectives according to the Euclidean distance, the L_2 norm.

The decision phase starts by presenting the efficient solution in the objective space defined by (6.1) to the DM. If the DM accepts the optimum solution after comparing it with the ideal vector, then the process ends; if it is not acceptable, then the DM must indicate which attribute(s) of the solution could be worsened or degraded so that the others could be improved. He must also indicate the maximum degradation possible before a satisfactory level of an attribute becomes unsatisfactory. This information imposes the following additional restraints on the problem before a new feasible set is generated:

$$Z_k(\underline{x}) \geq Z_k^1 - \Delta Z_k$$
$$Z_j(\underline{x}) \geq Z_j^1 \qquad\qquad j = 1, 2, ..., k-1, k+1, ..., q \qquad\qquad (6.4)$$

where $Z_k(\underline{x})$ is the satisfactory objective, ΔZ_k is the maximum degradation allowed in its achievement level, and the vector $[Z_1^1, ..., Z_q^1]$ is the solution in the objective space.

For the next iteration obviously $v_k = 0$, $W_k = 0$ and therefore the other normalising weights have to be recalculated. With the new W_j weights and the new feasible set as augmented by the additional restraints from (6.4) a new efficient solution is obtained, which is once again evaluated by the DM. This iterative process goes on until the DM is satisfied with a given solution. Benayoun et al. (1971) claim that their method converges to a solution in less than q iterations, q being the number of objectives; otherwise, the STEM method is not suitable for modelling the preferences of the DM.

To illustrate the functioning of the STEM method, our example from Chapter 1 has been modified and made a little more complex by introducing an additional crop and another objective, leading to the following general MOP model:

$$Eff\, Z(\underline{x}) = [Z_1(\underline{x}), Z_2(\underline{x}), Z_3(\underline{x})]$$

where

$$Z_1(\underline{x}) = 1000x_1 + 3000x_2 + 1500x_3 \qquad\qquad (6.5)$$
$$Z_2(\underline{x}) = 500x_1 + 200x_2 + 200x_3$$
$$Z_1(\underline{x}) = -6000x_1 - 8000x_2 - 3000x_3$$

subject to

$$
\begin{array}{rrrr}
x_1 + & x_2 + & x_3 \leq & 1000 \\
4000x_1 + 5000x_2 & + 2000x_3 \leq & 4{,}200{,}000 \\
-x_1 + & x_2 + & x_3 \leq & 0 \\
1000x_1 + 3000x_2 & + 1500x_3 \geq & 1{,}000{,}000 \\
500x_1 + & 200x_2 + & 200x_3 \geq & 350{,}000
\end{array}
$$

Recall that $Z_1(\underline{x})$ and $Z_2(\underline{x})$ represent, respectively, the value added and employment attributes. The third attribute, $Z_3(\underline{x})$, which has been included in the analysis now, represents the annual consumption of irrigation water that has to be minimised. The signs for the coefficients of $Z_3(\underline{x})$ have been reversed in order to establish the efficiency of all objectives in a maximisation sense. Thus the coefficient -6.000 means that crop A demands 6.000 m3/ha of water. The last two restraints of (6.5) have been specified to secure, respectively, a value added and a level of employment not less than £1,000,000 and 350,000 hours – the achievement levels of these attributes obtainable before the region was developed.

The satisfactory levels of $Z_1(\underline{x})$, $Z_2(\underline{x})$ and $Z_3(\underline{x})$ for a particular DM in the above problem are sought through the use of the STEM method. As pointed out earlier, the first step is to generate the pay-off matrix, by optimising the three objectives separately and then computing the value of every objective for each of the respective optimum solutions. This pay-off matrix is shown in Table 6.1.

Table 6.1 STEM pay-off matrix

	Value added (£)	Employment (hours)	Water (m3)
Value added	1,850,000	350,000	6,500,000
Employment	1,000,000	500,000	6,000,000
Water	1,000,000	350,000	4,363,560

By using the information from the pay-off matrix, it is possible to obtain the normalising weights W_js as below:

$$
v_1 = \left(\frac{1{,}850{,}000 - 1{,}000{,}000}{1{,}850{,}000} \right) \cdot \left(\frac{1}{\sqrt{1000^2 + 3000^2 + 1500^2}} \right) = 1.311 \times 10^{-4}
$$

$$
v^2 = \left(\frac{500{,}000 - 350{,}000}{500{,}000} \right) \cdot \left[\frac{1}{\sqrt{500^2 + 200^2 + 200^2}} \right] = 5.222 \times 10^{-4}
$$

$$
v_3 = \frac{6{,}500{,}000 - 4{,}363{,}500}{6{,}500{,}000} \cdot \frac{1}{\sqrt{6000^2 + 8000^2 + 3000^2}} = 0.315 \times 10^{-4}
$$

Therefore, from the v values we obtain:

$$W_1 = \frac{1.311}{6.848} = 0.191 \qquad W_2 = \frac{5.222}{6.848} = 0.763 \qquad W_3 = \frac{0.315}{6.848} = 0.046$$

Using the normalising weights W_js the following LP model is solved:

Mind d

subject to

$$0.191\,[1{,}850{,}000 - 1000x_1 - 3000x_2 - 1500x_3] \le d$$
$$0.763\,[500{,}000 - 500x_1 - 200x_2 - 200x_2] \le d \qquad (6.6)$$
$$0.046\,[6000x_1 + 8000x_2 + 3000x_3 - 4{,}363{,}560] \le d$$

$\underline{x} \in \underline{F}$ [technical constraints from model (6.5)]

The optimal solution for this problem is:

$$\underline{x}^1 = [x_1^1 = 671.96;\ x_2^1 = 195.58;\ x_3^1 = 132.46]$$

The image of this solution in the objectives space is:

$$Z^1 = [z_1^1 = 1{,}547{,}384;\ z_2^1 = 401{,}588;\ z_3^1 = 5{,}993{,}776]$$

Once the calculation phase of STEM has finished, the decision phase starts. The solution vector $\underline{Z}^1$ and the ideal vector, $\underline{Z}^*$, are presented to the DM for comparison. Supposing that the DM decides that z_1^1 is satisfactory, that is the level of value added achieved is sufficient, but the level of employment is too low and the consumption of water too high. Moreover, the DM feels that he could accept a value added of only 1,200,000, that is he accepts a degradation of z_1^1 attribute by 257,384, so that the level of employment is improved and water consumption is reduced. The DM will have local preferences with regard to the solution $\underline{Z}^1$, which are used by the analyst in determining how the model is re-run. The new normalising weights are:

$$W_1 = \frac{0}{5.537} = 0 \qquad W_2 = \frac{5.222}{5.537} = 0.94 \qquad W_3 = \frac{0.315}{5.537} = 0.06$$

and the new constraint set now is:

$x \in F$ [technical restraints from the model (6.5)]

and

$1000x_1 + 3000x_2 + 1500x_3 \geq 1,457,384 - 257,384 =$
$1,200,00 (6.7)$
$\quad 500x_1 + 200x_2 + 200x_3 \geq 401,588$
$\quad 6000x_1 + 8000x_2 + 3000x_3 \leq 5,993,766$

If we let F represent the new constraint set, in the next calculation phase of the STEM method the following linear programming model is solved:

Min d'

subject to

$0.94 \, [500,000 - 500x_1 - 200x_2 - 200x_3 \,] \leq d' \qquad (6.8)$
$0.06 \, [6000x_1 + 8000x_2 + 3000x_3 - 4,363,560] \leq d'$

and

$x \in F^1$

The optimum solution for this problem is:

$X^2 = [x_1^2 = 754.12; \, x_2^2 = 51.37; \, x_3^2 = 194.51]$

The image of this solution in the objective space is:

$Z^2 = [z_1^2 = 1,200,000; \, z_2^2 = 426,238; \, z_3^2 = 5,519,264]$

After evaluating Z^2 and comparing it with the ideal vector Z^* the DM may consider that the three objectives are at a satisfactory level of achievement and the algorithm terminates. If DM does not feel happy with this solution another iteration is carried out. To do that the DM would have to indicate if the objective of either employment or water consumption has achieved a satisfactory level and what is the maximum degradation possible for either of these. The main steps that comprise the STEM method are summarised in Figure 6.2.

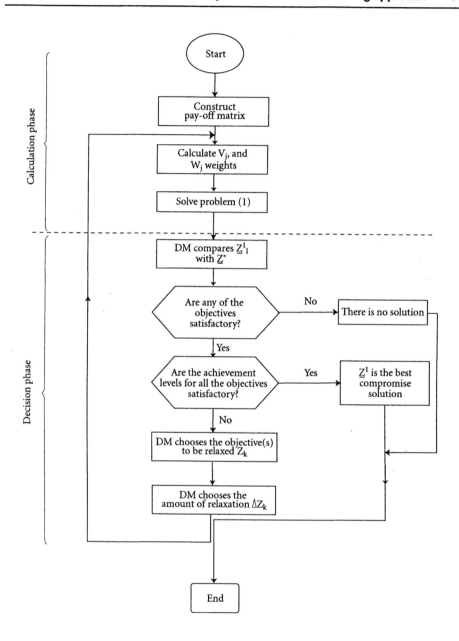

Figure 6.2 Main steps of the STEM method

The Zionts and Wallenius method

Zionts and Wallenius (1976) have proposed an interactive MCDM approach which is perhaps the most popular method for the second type of communication between the DM and the model explained in the first section of this chapter. This method, in common with all types of interactive approaches, has a calculation phase where the DM is not involved at all and a decision phase involving interaction with the DM. In the Zionts and Wallenius (hereafter ZW) method the calculation phase is rather complex while the decision phase is relatively easy. As the calculation phase of the ZW method is quite complex, we have refrained from explaining its mathematical structure first and have instead illustrated the mechanics involved straight away using the example from the last section.

The first step in the calculation phase of the ZW method is to solve an LP problem to optimise a linear composite function consisting of the objectives being considered in the model after having attached arbitrary weights to them (although usually the method starts by assigning equal weights to all the objectives). When the objectives are built into the matrix as equalities then the size of the constraint set is increased. These auxiliary restraints play an essential role in this method as they generate trade-off vectors that are presented to the DM. So, for our example, we start by solving the following linear programme:

$$\text{Max}\quad 0.33Z_1 + 0.33Z_2 - 0.33Z_3$$

subject to

$$
\begin{aligned}
Z_1 &= 1000x_1 + 3000x_2 + 1500x_3 \\
Z_2 &= 500x_1 + 200x_2 + 200x_3 \\
Z_3 &= 6000x_1 + 8000x_2 + 3000x_3
\end{aligned}
\qquad (6.9)
$$

and

$$\underline{x} \in F \,[\text{technical constraints from the model (6.5)}]$$

The objective function of (6.9) represents something like an arbitrary surrogate utility function for the DM where each objective has an equal weight. Once the above LP problem has been solved, the next step is to analyse that part of the optimal final simplex tableau, which corresponds to the basic rows Z_1, Z_2, and Z_3, and to the non-basic variables. This part of the final tableau for our example is reproduced in Table 6.2.

Table 6.2 Final simplex model for the first iteration of the Zionts–Wallenius method

Basis	x_1	x_3	Z_1	Basic variables Z_2	Z_3	...	x_2	Non-basic variables s_1	q_2
x_1									
x_3									
Z_1							-1500	1883	1.67
Z_2							0	0	-1
Z_3							-5000	1000	-10

In this tableau s_1 is the slack variable for the first constraint and q_2 is the surplus variable for the last constraint of model (6.5). We know that in a simplex tableau the coefficient of a non-basic variable against a row of the basis or the basic variable gives the amount by which the level of that basic variable will be reduced if a unit of the non-basic variable is introduced. Thus, the vector [-1500, 0, -5000] implies that if x_2 enters into the basis a unitary increase in the value of that variable will increase the value added by 1500, will not affect the level of employment and will increase the water consumption by 5000 m3. Therefore the following vectors can be interpreted as possible trade-offs that can be considered by the DM.

$$T_1 = [-1500, 0, -5000], T_2 = [1883, 0, 1000] \text{ and } T_3 = [1.67, -1, -10]$$

Now the non-basic variables have to be partitioned into efficient and non-efficient ones. In this context an efficient non-basic variable is one that when introduced into the basis leads to an adjacent extreme efficient point. To decide whether a certain non-basic variable is efficient or not, the ZW method uses a test as explained below with reference to the non-basic variable x_2. Thus, in order to ascertain whether or not that variable is efficient, the following linear programme has to be solved:

$$\text{Min } \phi = -1500A_1 + 5000A_3$$

subject to

$$
\begin{aligned}
1833A_1 \quad &- 1000A_3 \geq 0 \\
1.67A_1 - A_2 &+ 1000A_3 \geq 0 \\
A_1 + A_2 + \quad &A_3 = 1
\end{aligned}
\tag{6.10}
$$

where A_1, A_2, and A_3 represent the unknown weights to be attached to each of the attributes respectively in the composite objective function.

The objective function ϕ of (6.10) minimises the reduction in the utility for the DM (or maximises the increase in utility) when the non-basic variable x_2 enters the basis. As the two inequalities of (6.10) secure a decrease in utility when the other two non-basic variables enter into the basis, then if the optimum value of ϕ is negative (i.e. there is an increase in utility) the entry of the non-basic variable x_2 will lead to an efficient solution. For instance, it is easily checked that when $A_1 = 1$, $A_2 = A_3 = 0$, the constraints are satisfied and $\phi < 0$, therefore x_2 is an efficient variable.

Likewise to check the efficiency of the two other non-basic variables s_1 and q_2, the following two linear programmes problems must be solved:

$$\text{Min } \phi' = 1833\, A_1 - 1000 A_3$$

subject to

$$
\begin{aligned}
-1500\, A_1 \quad\quad + 5000 A_3 &\geq 0 \\
1.67 A_1 - A_2 + \quad 10 A_3 &\geq 0 \\
A_1 + A_2 + \quad\quad A_3 &= 1
\end{aligned}
\tag{6.11}
$$

and

$$\text{Min } \phi'' = 1.67\, A_1 - A_2 + 10 A_3$$

subject to

$$
\begin{aligned}
-1500\, A_1 \quad\quad + 5000 A_3 &\geq 0 \\
1833\, A_1 \quad\quad - 1000 A_3 &\geq 0 \\
A_1 + A_2 + \quad\quad A_3 &= 1
\end{aligned}
\tag{6.12}
$$

In both cases feasible solutions for which ϕ' and ϕ'' are negative are easy to find; therefore, the non-basic variables s_1 and q_2 are also efficient. Once the efficiency of the non-basic variables has been established the decision phase of the ZW method can be implemented. For that, the three trade-off vectors T_1, T_2 and T_3 are presented to the DM, asking him if they are desirable, not desirable or indifferent. Thus, for T_1 the DM would be asked: Do you accept an increase in value added of 1,500, accompanied by no change in the level of employment and an increase of 5000 m^3 in water consumption? (assuming that the DM is specially interested in the value added attribute and therefore considers the trade-off acceptable). If the answer is consistent with his preferences then the trade-off will add to his utility (that is it leads to a negative decrease in utility) and implies the following constraints:

$$-1500A_1 + 5000A_3 < -\beta \tag{6.13}$$

where β is a sufficiently small perturbation factor which in many applications equals 0.001, in order to impose a negative decrease in utility.

Let us assume now that the DM does not accept the other two trade-offs. As far as the DM is concerned both of these trade-off vectors imply a decrease in utility; hence the following constraints must be included in the problems:

$$1833A_1 - 1000A_3 > \beta \tag{6.14}$$

$$1.67A_1 - A_2 + 10A_3 > \beta \tag{6.15}$$

The values of A_1, A_2, and A_3 satisfying the constraints in (6.13) to (6.15) plus the normalising restraint $A_1 + A_2 + A_3 = 1$ represent sets of weights to be attached to the different attributes that are consistent with the preferences shown by the DM during the decision phase. In our case, $A_1 = 0.50$, $A_2 = 0.40$, $A_3 = 0.10$, is a possible set of weights meeting the above constraints. Therefore, an approximation of the DM's utility function obtained in the first iteration of the ZW method is: $U = 0.50Z_1 + 0.40Z_2 - 0.10Z_3$.

The second iteration of the ZW method starts with a new calculation phase. For that, the LP problem given by (6.9) is now solved for the new set of weights:

$$A_1 = 0.50, A_2 = 0.40, A_3 = 0.10$$

leading to the following extract of the final tableau (Table 6.3).

Table 6.3 Final simplex tableau for the second iteration of the Zionts–Wallenius method

Basis	x_1	x_2	x_3	Basic variables Z_1	Z_2	Z_3	...	Non-basic variables s_1	s_2	q_2
x_1										
x_2										
x_3										
Z_1								-1500	0.5	5
Z_2								0	0	-1
Z_3								-111.11	1.67	1.1

Here s_1 and q_2 are as defined before and s_2 corresponds to the slack variable of the second constraint of the model in (6.5). On testing the efficiency of the non-basic variables according to the method described above, s_2 and q_2 are found efficient, whereas the variable

S_1 is obviously a non-efficient one as it decreases the value added by £1500 and increases the consumption of water by 111.11 m3. There are, therefore, only two trade-off vectors to be considered by the DM, that is $T_1 = [0.5\ 0\ 1.67]$ and $T_2 = [5\ -1\ 1.1]$. Let us assume that the DM does not accept the trade-offs because they offer very little increase in employment and a small decrease of water consumption for a relatively large sacrifice in the value added. As no further attractive trade-off is found, the process is stopped and the solution obtained in the second iteration is regarded as the best approximation for the point of maximum utility. This solution is $x_1 = 500$, $x_2 = 400$, $x_3 = 100$, representing the added value of 1,850,000, and an employment level of 350,000 hours, and the water consumption level of 6,500,000 m3.

On the other hand, if in the second iteration at least one of the trade-off vectors was considered desirable by the DM, then the corresponding restraints are added to the constraint set (6.13) to (6.15) and, obtaining from this restraint set a new set of weights that is compatible with the DM's preferences. With those weights the process would start again until the iteration at which no non-basic variable is efficient and/or no trade-off vector interests the DM. The broad outline of the mechanics of this method is shown in Figure 6.3.

The ZW method claims that since there are a finite number of extreme points, and that at each iteration of the algorithm at least one extreme point is eliminated, the convergence to an acceptable point (solution) is guaranteed. The underlying theory of the ZW method assumes that we are dealing with q concave objective functions of the decision variables and that the utility function of the DM is a linear combination of the objectives being considered in the model, but the weights of these are not known explicitly. Zionts and Wallenius (1983) have developed their method assuming the existence of an unknown pseudo-concave utility function satisfying certain general properties. This way some technical and conceptual problems present in the original ZW method have been overcome but the computational needs have increased considerably.

Interactive multiple goal programming

This section presents an interactive approach for using goal programming (GP) as proposed by Nijkamp and Spronk (1980) and by Spronk (1981). This is referred to as interactive multiple goal programming (IMGP) and falls within the third scheme of communication defined in the first section of this chapter. The main aspects of IMGP are explained by using the orchard-planning model from the previous chapters. However, now we assume that the DM is interested in three attributes: (1) the NPV or the investment in plantation; (2) the amount of hired casual labour for pruning and harvesting; and (3) to use 1000 hours of own tractor capacity. The DM is aiming to utilise his own tractor capacity fully; that is, he does not want either over-achievement (meaning hiring of tractor hours) or under-achievement (implying leaving his own tractors idle). In the context of our example, it is assumed that over-utilisation is four times more important than under-utilisation. The mathematical representation of the three attributes and the constraint set of our problem is given by:

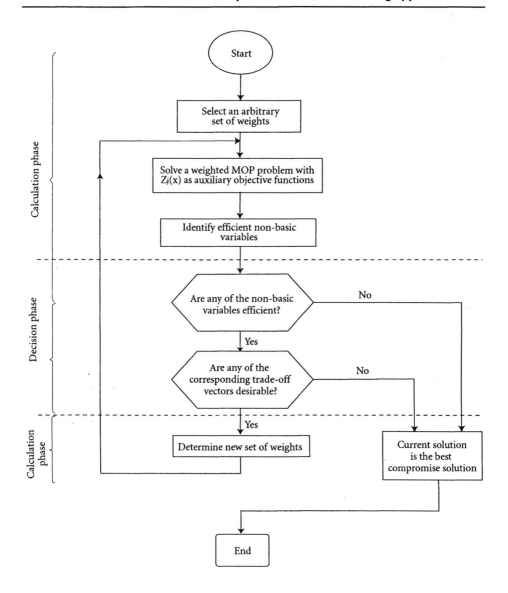

Figure 6.3 Main steps of the Zionts-Wallenius method

Attributes

(NPV)	$6250x_1 + 5000x_2$
(Casual labour)	$520x_1 + 630x_2$
(Use of own tractors)	$n_1 + 4p_1$

(6.16)

Constraints and goals

$$1375x_1 + 1025x_2 \leq 36{,}000$$
$$x_1 + x_2 \geq 10$$
$$35x_1 + 35x_2 + n_1 - p_1 = 1{,}000$$

The general purpose of IMGP is to obtain the targets or aspiration levels from the DM interactively. The first step is to derive what Spronk (1981) calls the *potency matrix*. The elements of the first row of this matrix are the ideal or optimum values, while the second row consists of the anti-ideal values of the attributes under consideration. On optimising the problem in (6.16), in turn for each attribute, we obtain the following potency matrix:

$$P_1 = \begin{bmatrix} 175{,}600 & 5{,}200 & 0 \\ 62{,}500 & 22{,}126 & 4{,}917 \end{bmatrix}$$

Evidently each column of the potency matrix refers to a particular attribute and as such the elements of a column represents the range within which it is possible to define the targets of a particular attribute. It is not possible to fix a target higher than the ideal value; nor it is rational to fix a target lower than the anti-ideal value. Hence the target for the NPV attribute is between £62,500 and £175,600.

For each attribute the DM may have defined tentative targets. Let us assume that this vector of tentative targets is $T = [150{,}000 \; 8{,}000 \; 100]$. Given the potency matrix and the DM's vector of targets, the next step is to present to him the anti-ideal or pessimistic vector as the first proposal solution, S_1:

$$S_1 = [62{,}500 \quad 22{,}126 \quad 4{,}917]$$

If the DM considers the proposed solution S_1 as satisfactory, an unlikely occurrence, the process ends. If the solution is not acceptable, then the DM must indicate the attribute that should be improved first. Let us assume that it is desirable to improve the casual labour attribute first. Then, a new proposal solution $\hat{S}_2$ is generated by substituting the second component of S_1 by the second element of the vector of targets T. The second proposed solution then is:

$$\hat{S}_2 = [62,500 \quad 8,000 \quad 4,917]$$

Next we calculate another potency matrix associated with the new solution $\hat{S}_2$, by augmenting the constraint set of (6.16) with the additional restraint of $520x_1 + 630x_2 \leq 8,000$. The new potency matrix is:

$$\hat{P}_2 = \begin{bmatrix} 96,187 & 5,200 & 461 \\ 62,500 & 8,000 & 4,917 \end{bmatrix}$$

The shifts in the potency matrix can be interpreted as the opportunity costs for reaching the new proposed solution. Faced with information in S_1 and $\hat{S}_2$ and in P_1 and $\hat{P}_2$, the DM must decide if the sacrifice or opportunity cost in moving from one proposed solution to another is justified. Suppose he regards that the improvement in the casual labour attribute is accompanied by rather too much degradation of the NPV attribute; therefore, we must set a less ambitious target for the casual labour attribute. Spronk (1981) suggests that this new value should be chosen exactly halfway between the corresponding value in S_1 and in the rejected solution $\hat{S}_2$, that is $(22,126 + 8,000)/2 = 15,063$ for our case. So, the second proposed solution is:

$$\underline{\hat{S}}_2 = [62,500 \quad 15,063 \quad 4,917]$$

In order to obtain the new potency matrix $\underline{\hat{P}}_2$ the right-hand side of the additional restraint is substituted by the new target, that is the restraint $520x_1 + 630x_2 \leq 15,063$ is added to the constraint set of (6.16); therefore, the revised second potency matrix is:

$$\underline{\hat{P}}_2 = \begin{bmatrix} 165,625 & 5,200 & 30 \\ 62,500 & 15,063 & 4,917 \end{bmatrix}$$

Assuming that the DM considers that the shifts in the potency matrix are justified, we can regard $S_2 = \underline{\hat{S}}_2$ and $P_2 = \underline{\hat{P}}_2$; note that S_2 and P_2 are the solution and the potency matrix that have been obtained after the first improvement in S_1, the previous solution, has been accepted by the DM.

After the above stage, the DM should now indicate the second attribute to be improved. Assuming that he chooses the NPV attribute, the new $\hat{S}_3$ proposal solution is obtained by substituting the first component of S_2 by the first element in the tentative target vector T. So we have:

$$\hat{\underline{S}}_3 = [150{,}000 \quad 5{,}063 \quad 4{,}197]$$

By augmenting the set of constraints (6.16) by the new restraint $6{,}250x_1 + 5{,}000x_2 \geq 150{,}000$, the following new potency matrix is obtained:

$$\hat{\underline{P}}_3 = \begin{bmatrix} 165{,}625 & 12{,}840 & 30 \\ 150{,}000 & 15{,}063 & 4{,}917 \end{bmatrix}$$

If the DM accepts the new shifts in the potency matrix, then we make $S_3 = \hat{\underline{S}}_3$ and $P_3 = \hat{\underline{P}}_3$. The next step is to seek an improvement in the attribute of using own tractors. Following the procedure outlined in the preceding steps, we obtain the new proposal solution as:

$$\hat{\underline{S}}_4 = [150{,}000 \quad 15{,}063 \quad 100]$$

After including the new restraint of $n_1 + 4p_1 \leq 100$ to the constraint set (6.16), the new potency matrix is:

$$\hat{\underline{P}}_4 = \begin{bmatrix} 165{,}625 & 14{,}312 & 30 \\ 150{,}000 & 15{,}063 & 100 \end{bmatrix}$$

The DM may now regard the shifts in the potency matrix as satisfactory and find the proposal solution $\hat{\underline{S}}_4$ as acceptable. In that case the interactive process stops. On the other hand, if the DM finds the potency matrix shifts satisfactory, but is not happy with the proposal solution $\hat{\underline{S}}_4$ then he may add further restraints to the constraint set (6.16) in order to generate another potency matrix and therefore a new proposal solution.

The interactive process will continue until a satisfactory proposal solution for the DM has been found, and also when the two rows of the potency matrix are equal. The second requirement of the two rows of the potency matrix being equal implies that the ideal and the anti-ideals of each attribute coincide so that one cannot proceed with the process. However, the rows of the final potency matrix represent the vector of targets obtained from the DM interactively.

It should be noted that the set of targets required from the DM for use by IMGP must lie within the interval defined by the ideal and anti-ideal values of each attribute being considered. Thus for the jth attribute, the DM should provide the following targets:

$$g_{\cdot j} < g_{j1} < g_{j2} \cdots g_{jk} < g_j^*$$

where g_j^* and $g_{\cdot j}$ are the ideal and the anti-ideal values respectively. A summary of the steps that need to be carried out in this method is given in Figure 6.4.

As a final remark, it is not absolutely necessary to have an initial vector of tentative targets set by the DM in order to start IMGP. The method can work without that vector. At each step of the process, the DM can indicate the attribute to be improved by fixing its target. However, if the vector or vectors of targets are provided by the DM *a priori* at the beginning, then the conduct of the interactive process is relatively easy; as the DM is required only to interact with the model to decide if he accepts, or not, the shifts in the potency matrix.

An assessment of interactive MCDM approaches

Even though we have covered only three methods, they do represent the different interactive techniques sufficiently well to enable us to assess to the general possibilities in this area of MCDM modelling.

Starting with the STEM method, the main advantage lies in its operational simplicity. It is only necessary to solve a maximum of q linear programming problems, where q is the number of objectives. It does not require a previous generation of the efficient set. At each calculation phase only one efficient solution is obtained. Further, the lack of restrictive assumptions about a DM's absolute preferences and the shape of his utility function is a practical advantage. It only requires the DM to define, partially, his local preferences at each decision phase of the process, to obtain an approximation of the point of his maximum utility.

The main problem or weakness, however, is perhaps in the rather intensive interaction with the DM, requiring information from him, at each decision phase, not only on the objective with a satisfactory achievement level, but also on the precise amount by which it could degrade in order to affect improvement(s) in the other attributes. Many decision makers may not be in a position to provide that information consistently; therefore, the accuracy of the solution depends mainly on the capacity of the DM to answer such difficult questions correctly and consistently.

We should note that the STEM method is primarily a reduced feasible region method; therefore, the normalising system or the metric used to generate the efficient solutions can be modified without changing its basic purpose; that is, to reduce the size of the feasible set through imposing constraints obtained interactively from the DM. However, if the metric used is L_1 and not L_∞, then we are actually using the weighting method (see Chapter 5); therefore, at each decision phase, a restricted set consisting of only extreme efficient points is presented to the DM for evaluation. However, using other metrics, such as the L_∞, the efficient points evaluated include both extreme and interior points, making this set larger.

The ZW method lies at the other end of spectrum to the STEM approach. The calculation phase is very complex, mainly due to the procedure followed in determining the non-basic efficient solutions. Although the problem being analysed would have few objectives if the constraint set is large, even then the number of non-basic variables to be tested for efficiency is considerable. This implies having to solve an important number of small auxiliary LP problems. Zionts and Wallenius (1980) have presented simplified methods for finding the

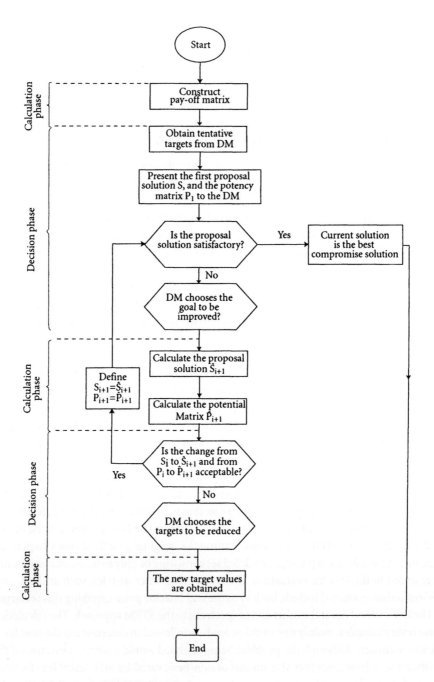

Figure 6.4 Main steps of the IMGP method

subset of efficient vectors of an arbitrary set of vectors to mitigate that problem; however, the computational difficulties of the ZW method remain its weak point.

Another possible weakness of the ZW method lies in the strong assumption that the DM's utility function is an unknown linear function of his objectives. Zionts (1983, pp. 417–418) argues that this difficulty is apparent because the ZW method uses that function only to identify good alternatives rather than using it as a proper utility function. The ZW method also operates as the weighting method; therefore, in each iteration only extreme efficient points are considered. In other words, we obtain the extreme efficient point of maximum utility interactively. But we know that in many instances an interior point can be better in terms of utility than an extreme point.

The decision phase of this method is, however, attractive as the questions raised can be easily understood and answered by the DM. The questions require only a partial knowledge of the local trade-offs of the DM. However, the assumption that the DM makes all his choices such that they are consistent with an unknown and implicit utility function should be pointed out. Inconsistent answers given by the DM, particularly when he is dealing a complex decision-making situation involving a comparison of many trade-offs, can produce disappointing results. For instance, if in the second decision phase of the example in the section on ZW the DM would have accepted the trade-off vector [0.5 0 1.67], then the following restraint would have been included in the auxiliary LP problem:

$$0.5A_1 - 1.67A_3 \leq - \beta \qquad\qquad (6.17)$$

Obviously the constraints (6.13) and (6.17) are incompatible. In other words, there is no utility function that is compatible with the preferences shown by the DM. It is likely that many decision makers produce similarly inconsistent answers under intensive questioning.

The appeal of IMGP lies in the relative ease with which its calculation phase is implemented and the simplicity of information required from the DM. This is especially true when it is compared with another interactive GP approach that has been proposed by Dyer (1972). IMGP is an easily implemented 'trial-and error' procedure, which is both its main advantage and the disadvantage.

The questions that the DM is asked are easy to answer. Similarly the computations involved in each calculation phase are also easy. However, the main handicap is the number of questions, depending upon the number of decision phases that can be presented to the DM. Too large a number of questions increases the likelihood of inconsistent answers. We must also note that IMGP, more than any other interactive method, is a framework or a general approach to tackling decision-making problems involving multiple goals interactively. Several variants of this approach can be found in Spronk (1981).

To summarise, the main advantages of the interactive approach within a MCDM framework are:

1 it represents a learning process for the DM, permitting him to better understand the system being analysed;

2 the information required involves only the local preferences of a DM, that is, his attitude towards a certain solution or with respect to a certain set of trade-offs; and

3 in general the assumptions underlying an interactive method are much less restrictive than those required for a non-interactive technique.

What must be stressed about the interactive approach, however, is the fact that in this process all the entities play their respective roles as those roles ought to be played. The decision maker takes the decisions while the analyst analyses the problem and the model is used to make the necessary calculations. The interactive approach, however, has its difficulties and they are:

1 the effort and involvement required from the DM in using the model is considerably more when compared with non-interactive methods;

2 the assumption that the DM makes all his decisions consistently, particularly when inconsistencies can be common;

3   in some interactive methods the number of iterations necessary to converge to a solution can be very important – thus, the ZW method only guarantees that the number of iterations will be less than the number of extreme efficient points; and,

4 from some experiments it is possible to deduce that DM has little confidence in some interactive methods, finding them more difficult to use than some 'trial-and-error' methods – thus, an experiment undertaken by Wallenius (1975b) shows that a sample of Finnish business students and managers preferred a 'trial-and-error' approach to the STEM and Geoffrion et al. methods.

To conclude this chapter we offer two remarks. First, the interactive approach to multiple criteria decision-making represents an attractive and realistic way of solving these problems, even though intensive research is still required to develop theoretically and operationally sound methods. Second, the choice of a specific interactive method depends upon not only on the problem situation in hand, but also on the characteristics of the DM such as his cultural background, education, mathematical abilities, training and so on.

Suggestions for further reading

The literature on the interactive MCDM approach is very extensive. For a state-of-the-art review see Hwang et al. (1979, pp. 102–226). A good technical treatment of the main methods can be seen in Steuer (1986, pp. 361–446). The book of readings edited by Grauer and Wierzbicki (1984) is an excellent presentation of the advances undertaken in the 1970s, even though the level of mathematical training required to follow the text is high. A comprehensive survey of these techniques can be found in Olson (1992). A unified approach embracing several interactive methods can be seen in Gardiner and Steuer (1994).

The STEM method has been modified and extended in several ways. Fichefet (1974) merges it with GP, producing an interesting interactive method known as GPSTEM. Teghem and Kunsch (1985) and Teghem et al. (1986) have presented the STRANGE method, which is basically an extension of the STEM approach to deal with uncertainty. Other extensions of the STEM method can be found in: Dinkelback and Iserman (1980); Michalowski and Piotrowski (1983); Michalowski and Zolkiewski (1983); and, Slowinski and Warczyinski (1984).

Pedagogical presentations and overviews of the ZW method include: Wallenius (1975a); Wallenius and Zionts (1977); Samblanchx et al. (1982); Zionts (1983). For theoretical extensions and improvements consult: White (1980); Zionts (1981); Zionts and Wallenius (1983); Stewart (1984, 1986); Michalowski (1985). Some notable applications of the ZW method to real problems include Wallenius et al. (1978) who apply it to macroeconomic planning in Finland and the work of Zionts and Deshpande (1978, 1981) for energy planning problems, Mathiesen (1981) for fisheries management and Lara and Romero (1992) for a livestock ration formulation problem.

Close relatives of the IMGP method are the reference point optimisation method developed by Wierzbicki (1982) and the method of Weistroffer (1984). Spronk and Veeneklas (1982) have applied the IMGP method to macroeconomic planning and Spronk and Zambruno (1985) have also used it for bank portfolio selection. For other interactive GP methods that have been proposed see, among others: Dyer (1972); Monarchi et al. (1976); Franz and Lee (1981); Ignizio (1981); Masud and Hwang (1981); Spronk and Telgen (1981); Weistroffer (1982); Tabucanon and Mukyangkoon (1985).

Gabbani and Magazine (1986), Marcotte and Soland (1986) and Ramesh et al. (1986) have adapted interactive MCDM methods to solve integer programming problems. Choo and Atkins (1980) have presented an interactive method where the objectives are fractional.

Other interactive approaches are the Belenson and Kapur (1973) method where the weights of the MOP problem are interactively elicited by applying LP to the solution of a two-person-zero sum game; the surrogate worth trade-off method of Haimes and Hall (1974) based on the constraint method and which has been applied extensively to water resources planning; the method of Steuer (1977) based on interval criterion weights; the sequential proxy optimisation technique of Sakawa (1982); the interactive weighted Tchebycheff procedure of Steuer and Choo (1983); and the visual interactive method of Korhonen and Laakso (1986). A critical assessment and an analysis of the foundations of these interactive approaches can be seen in White (1983) and French (1984).

The pioneering applications of the interactive MCDM approach in agriculture are that of (1) de Wit et al. (1988) for a regional agricultural planning exercise and (2) Berbel et al. (1991) who apply the STEM method to the planning of an agribusiness company. However the number of applications of the interactive MCDM approach in agricultural decision problems is still low. This is intriguing because most of the actual decision-making problems in agriculture seem to possess characteristics that call for analysis within this framework. This is particularly striking when one looks at the current work in the neighbouring field of natural resources management, where considerable effort has been devoted to solve decision problems within an interactive framework. Among these applications are the papers by: Walker (1985) and Liu and Davis (1995) in forestry; Mathiesen (1981) and Stewart (1988) in fisheries; Sakawa (1984) and Greis et al. (1983) in water resources.

Finally, the analytic hierarchy process (AHP) method introduced by Saaty (1977, 1980) should be mentioned as belonging the set of the general interactive MCDM methods. Within this approach the DM is expected to provide information on the relative importance of the criteria involved in the decision-making process through a pair-wise comparison of the criteria. A set of weights that reflect the DM's responses to the above questions are then extracted from this information. In the last ten years the practical use of AHP has been impressive. Forman and Gass (2001) have recently provided a state-of-the-art introduction to the foundations of the AHP, whilst Alphonce (1997) has explored the possibilities of its application in agriculture.

Chapter seven

Risk and uncertainty and the multiple criteria decision-making techniques

This chapter[1] demonstrates how the usual methods of modelling risk and uncertainty, that are inherent in agricultural decision-making, can be incorporated within the multiple criteria decision-making (MCDM) framework. It is also shown that the traditional risk and uncertainty analysis, by its very nature, is multiobjective analysis involving two objectives: profit and a measure of its variability. Treating the risk and uncertainty models as particular cases of the MCDM paradigm has therefore both theoretical and practical advantages. It is possible to create a 'hybrid' of the Markowitzean and MOTAD approaches with compromise programming to obtain compromise sets, which lie closest to an ideal point defined in terms of an acceptable level of risk. Similarly, the game-theoretic approach can be extended to what may be called compromise games enabling us to explain the behaviour of the decision maker (DM), by considering a set of conflicting criteria rather than relying on the naïve assumption of a single criterion optimisation by decision makers.

The distinction between risk and uncertainty as proposed by Knight (1921) has proved useful over time and is, therefore, maintained in this chapter. Another point worth mentioning is that generally the models that deal with uncertainty are of game theoretic type and those that incorporate risk are based on mathematical programming techniques.

Risk programming techniques in agricultural planning within the MCDM framework

The first attempts at developing risk programming models for agricultural planning were inspired by the method proposed by Markowitz (1952). In using the Markowitzean approach, the risk of an agricultural enterprise is measured by the variability of its returns using variance as the index. Low-risk enterprises have relatively small variance which means that their returns are concentrated round the mean value; whereas high-risk enterprises have relatively large variance with their returns dispersed round the mean value. Once the risk associated with each enterprise has been established in this manner, the Markowitzean concept of efficiency is used. A 'portfolio' or a mixture of agricultural enterprises is efficient if it has minimum variance for a given level of income or it has a maximum income for a given level of variance.

The next step in the analysis is to generate this efficient set which is usually done through a parametric quadratic programming (QP) model; where the total variance of the plan, that is a mixture of enterprises, is the objective function to be minimised. The constraint set of the problem includes an additional restraint measuring the expected income of the plan. Parameterising of the right-hand side value of that restraint provides the Markowitzean efficient set. The first application of this method to an agricultural problem is due to Freund (1956).

On closer examination of the Markowitzean approach, it becomes obvious that it is akin to a multiobjective programming (MOP) model with two objectives: the expected income and its variance. Further, the efficient set is defined in terms that are similar to MOP's definition of efficiency, and the mechanics of its generation are similar to the constraint method (see Chapter 4).

Hazell (1971) has demonstrated that the minimisation of variance is equivalent to minimising mean absolute deviations (MOTAD), thereby enabling us to deal with an ordinary LP model rather than solve a QP problem. Hazell's approach defines the efficiency of solutions in terms of expected income and mean absolute deviation associated with the solutions; and once again, it is also an MOP model with two objectives. Moreover, as the minimisation of the mean absolute deviation implies minimising the sum of the deviational variables measuring under- and over-achievements with respect to a null deviation for every period considered in the model, then the MOTAD approach is in fact a goal programming (GP) problem (Romero and Rehman 1985a).

Similarly, the main variants of the Markowitz and MOTAD approaches can be regarded as particular cases of the structure embodied in the MCDM paradigm. A relatively recent and promising method of risk programming, called target MOTAD (Tauer 1983) is effectively a hybrid of MOP and GP, whereby a target level of income T, allowing a deviation y_r below T for a state of nature or the rth period is determined. Two objectives – maximisation of the expected income, and minimisation of the aggregated deviations measuring the total under-achievement below the target T – are considered. The efficient set is generated following a constraint method approach.

To develop the links between the MCDM techniques and the usual approaches for modelling risk and uncertainty, let us adopt the data used by Hazell (1971) to introduce his MOTAD model as shown in Table 7.1. This table shows a time series of gross margins for the four vegetable enterprises considered in the planning problem.

Table 7.1 Gross margin ($ per acre) for Hazell's example farm

Crops/year	t_1	t_2	t_3	t_4	t_5	t_6	Average gross margins ($)
Carrots (x_1)	292	179	114	247	426	259	253
Celery (x_2)	-128	560	648	544	182	850	443
Cucumbers (x_3)	420	187	366	249	322	159	284
Peppers (x_4)	579	639	379	924	5	569	516

Source: Hazell 1971, p.60

Hazell's example can be re-formulated as a bi-criteria LP model as shown in Table 7.2, shown after Table 7.3. The first three rows of that table are the technical constraints of the problem. The rows (4) – (9) represent the typical MOTAD constraints, which deal with the under- and over-achievement with respect to a null deviation for the average gross margin during each of the six years. The row (10) represents the minimisation of the sum of the absolute values of the gross margin deviations and row (11) represents the maximisation of the expected gross margin.

Hazell minimised the objective function Z_1 (i.e. the mean absolute gross margin deviation) while treating the other objective Z_2 (i.e. the expected gross margin) as a parametric restraint. Using a parametric LP code Hazell determined the 'change-of-basis' solutions as presented in the first five rows of Table 7.3. These points when interpreted within an MCDM framework are the efficient extreme points that form the efficient or Pareto optimal set.

Table 7.3 Feasible efficient extreme points and the associated cropping patterns for Hazell's example farm

	Objective functions		Decision variables			
Efficient extreme points	Z_1 – Mean absolute gross margin deviation ($)	Z_2 – Expected gross margin ($)	Carrots x_1 (acres)	Celery x_1 (acres)	Cucumbers x_1 (acres)	Peppers x_1 (acres)
$L_1 \rightarrow A$	2,753	62,769	72.26	26.80	83.92	17.02
B	9,301	73,574	32.85	28.03	81.64	57.48
C	12,533	77,329	19.15	28.46	80.85	71.54
D	12,787	77,529	16.59	26.80	83.41	73.20
E	13,479	77,996	-	27.45	100	72.55
$L_\infty \rightarrow S$	2,727	62,180	71.58	26.55	83.13	16.86

The points (solutions) given in Table 7.3 have been plotted in the objectives space in Figure. 7.1 to provide a trade-off or transformation curve between expected gross margin and risk measured as the mean absolute gross margin deviation. The actual values of the trade-offs (i.e. the opportunity costs) between the two objectives under consideration can be viewed as the slopes of the straight lines connecting the extreme efficient points of Figure. 7.1.

One of the advantages of treating the traditional risk programming methods as MCDM models is a practical one. As, in fact, both Markowitz's and Hazell's approaches are bi-criteria models then a precise efficient set can be generated by using the NISE method (see Chapter 4), obtained by an iterative use of the ordinary LP simplex algorithm.

Table 7.2 MOTAD model for Hazell's example farm formulated as a bicriteria linear programming problem

Real activities				Deviational variables (dollars)															
Carrots x_1	Celery x_1	Cucumbers x_1	Peppers x_1	n_1	p_1	n_2	p_2	n_3	p_3	n_4	p_4	n_5	p_5	n_6	p_6				
1	1	1	1													≤	200	Crop land (acres)	(1)
25	36	27	87													≤	10,000	Labour (hours)	(2)
-1	1	-1	1													≤	0	Rotational and market outlet restraints	(3)
39	-571	136	63	1	-1											=	0	Gross margin deviation ($) $- t_1$	(4)
-74	117	-97	123			1	-1									=	0	Gross margin deviation ($) $- t_2$	(5)
-139	205	82	-137					1	-1							=	0	Gross margin deviation ($) $- t_3$	(6)
-6	101	-35	408							1	-1					=	0	Gross margin deviation ($) $- t_4$	(7)
173	-261	38	-511									1	-1			=	0	Gross margin deviation ($) $- t_5$	(8)
6	407	-125	53											1	-1	=	0	Gross margin deviation ($) $- t_6$	(9)
				1	1	1	1	1	1	1	1	1	1	1	1		Z_1	Minimise the sum of the absolute values of the gross margin deviations ($)	(10)
253	443	284	516														Z_2	Maximise expected gross margin ($)	(11)

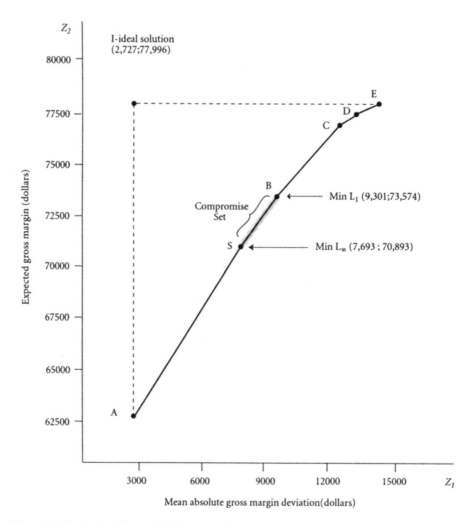

Figure 7.1 The trade-off curve for the expected gross margin and the mean absolute gross margin deviation

Compromise-risk programming

The methods proposed by Markowitz and Hazell do not provide any means of selecting the optimum solution from the efficient set once it has been generated. This problem, however, can be mitigated to a considerable extent if the risk programming models are grafted on to compromise programming (CP); such a compromise-risk programming is illustrated below by using, once again, Hazell's farm planning example.

The first step is to derive the pay-off matrix by optimising the two objectives one after the other, as explained in Chapter 4; thus the following pay-off matrix is obtained for Hazell's example.

	Mean absolute gross margin deviation	Expected gross margin
Mean absolute gross margin deviation Z_1	2,753	62,769
Expected gross margin Z_2	13,479	77,996

For these two solutions the corresponding cropping patterns are found in rows 1 and 5 (points A and E) of Table 7.3 respectively. The elements of the main diagonal of the pay-off matrix is the ideal vector for our problem; that is, to achieve a gross margin of 77,996 dollars with a mean absolute deviation of 2,753 dollars. The other elements of the matrix, that is the vector (13,479 62,769) represent the anti-ideal or the nadir vector.

The next step in the CP approach is to establish the best-compromise solution for the L_1 metric (i.e. for $p = 1$). This is done, as is the case in CP, by solving the following LP problem where the two objectives have been normalised according to the range between the ideal and the anti-ideal values (see Chapter 5).

$$\text{Min } L_1 = W_1 \frac{\frac{1}{6}\sum_{i=1}^{6} (n_i + p_i) - 2{,}753}{13{,}479 - 2{,}753} + W_2 \frac{77{,}996 - 253x_1 - 443x_2 - 284x_3 - 516x_4}{77{,}996 - 62769}$$

subject to   (7.1)

$\underline{x} \in F$ [constraints (1) to (9) from Table 7.2]

The optimum solution of the above LP problem for $W_1 = W_2$, that is, when the two attributes are equally important, is given by point B in the transformation curve of Figure 7.1. The point B therefore is the best-compromise solution; that is, B is the efficient point closest to the ideal point. As explained in Chapter 4, the other bound of the compromise set, corresponding to the L_∞ metric, is obtained by solving the following LP problem:

Min $L_\infty = d$

subject to

$$W_1 \frac{\frac{1}{6}\Sigma\,(n_i + p_i) - 2{,}753}{13{,}479 - 2{,}753} \leq d$$

$$W_2 \frac{77{,}996 - 253x_1 - 443x_2 - 284x_3 - 526x_4}{77{,}996 - 62{,}769} \leq d \tag{7.2}$$

$\underline{x} \in \underline{F}$ [Constraints (1) to (9) from Table 7.2]

The optimum solution of this LP problem, assuming $W_1 = W_2$, is given by point S for the transformation curve of Figure 7.1. The precise values of the objectives for this point and its corresponding cropping patterns are shown in the last row of Table 7.3. For the situation when $W_1 = W_2$ the segment SB is therefore the compromise set for our problem, representing the rational choices when the DM attaches equal importance to the discrepancies between the two objectives and their ideal values.

It is interesting to note that minimising the objective in (7.1) is the same as maximising the following function:

$$(253x_1 + 443x_2 + 284x_3 + 516x_4) - \frac{77{,}996 - 62769}{13{,}479 - 2{,}753}\frac{W_1}{W_2}\frac{1}{6}\sum_{i=1}^{6}(n_i + p_i) = \tag{7.3}$$

$$(253x_1 + 443x_2 + 284x_3 + 516x_4) - 1.42\frac{W_1}{W_2}\frac{1}{6}\sum_{i=1}^{6}(n_i + p_i)$$

But expression (7.3) maximises the utility function of the DM when that function is linear and additive for the two attributes under consideration and $1.42\ {}^{w_1}/_{w_2}$ is the risk-aversion coefficient.[2] This leads us to the following observations:

1 The bound of the compromise set corresponding to the metric L_1 is the point of maximum utility for the restrictive linear additive utility function; therefore, when in a conventional risk programming model the chosen utility function is linear and additive, it is implicitly assumed that the L_1 metric reflects the actual attitude of the DM towards the discrepancies between objectives and their ideal values. Obviously that assumption is very strong and restrictive but can be relaxed considerably to good advantage with the compromise-risk programming approach that has been proposed above.

2 It is not incorrect to say, at least intuitively, that for any conceivable form of the utility function (additive, multiplicative and so on) the point of maximum utility will belong to the compromise set. It is therefore quite legitimate to state that common use of linear and additive utility functions is only a particular case of Compromise-Risk Programming.

3 As the risk-aversion coefficient for the example is 1.42 $^{w_1}/_{w_2}$, then given two equally important attributes the implied value of that coefficient is 1.42. If the attribute of expected gross margin is twice as important as the other one, then the value of the risk-aversion coefficient decreases to 0.71, and so on. Given this interpretation of the weights, it would be worthwhile to examine how stable the compromise set is for different values of the risk-aversion coefficient as it takes on the values of 0.67, 0.071 and 142 respectively. However, the best-compromise point B is more unstable. For instance, that point moves to C or A when the coefficient takes on the values of 1.14 and 1.92 respectively.

To conclude, it has to be said that any variant of the MOTAD and Markowitz models, or the target MOTAD (Tauer 1983) approach to risk programming, can and should be accommodated within a compromise-risk programming structure. For target MOTAD, a compromise set for the attributes of expected gross margin and total aggregated deviations between gross margin and the target value specified can be obtained wherever the utility function of the DM achieves its maximum value, regardless of its shape.

Game theory models and the MCDM framework

Game theory models are one of the main conventional approaches to agricultural decision-making under uncertainty and consist of games against nature. The primary features of these games are:

1 a DM who is regarded as the only rational player of the game;

2 a set of n points representing the possible actions or strategies to be followed by the DM (e.g. the crop activities in a farm planning problem);

3 a set of m different possible states of nature representing the uncertainties within which the DM has to operate; and

4 an '$n \times m$' matrix whose r_{ij} elements represent the outcome of the game when the DM chooses the ith strategy in face of the jth state.

The purpose of a game theoretic model is to find a pure or mixed strategy that optimises the wishes and aspirations of the DM. This approach was introduced to agricultural decision-making by McInerney (1967, 1969).

A number of criteria have been used to represent the aspirations of the DM in the game theoretic models of agricultural planning. The most prominent amongst these are:

1 *The maximin or Wald criterion.* It assumes that DM looks for a strategy that maximises the outcome that can be achieved in the worst state of nature. In other words, a maximum minimum outcome is assured.

2 *The minimax regret or Savage criterion.* The first step in this criterion is the formulation of a pay-off matrix giving a 'regret' matrix. The elements of this matrix represent the

difference between the outcome actually achieved and the maximum outcome the DM could have achieved had he known the precise state of nature that would have prevailed. From the 'regret' matrix the Savage criterion looks for a strategy which minimises the largest possible 'regret' that any state of nature can produce.

3 *The benefit criterion.* Agrawal and Heady (1968) have argued that Wald's criterion is pessimistic, leading to conservative solutions, whereas Savage's is optimistic, leading to risky solutions. They offer a middle-of-the-road idea by the name of 'benefit' criterion that combines the properties of the other two.

In the Agrawal and Heady approach a 'benefit' matrix is formulated first where the elements represent the difference between the outcome actually achieved and the minimum that the DM could have obtained under the worst state of nature. From the 'benefit' matrix a strategy which maximises the minimum possible benefit under any state of nature is chosen.

Here again the traditional emphasis on using a single criterion as reflecting the behaviour of the DM seriously limits the applicability and relevance of the game theoretic framework to actual decision-making. We take the view that the actual behaviour of a DM is better approximated by a mixture of several criteria; because at any point in time within the game theoretic framework, a DM may be interested in combining, to a lesser or greater extent, the desires to maximise the minimum outcome, minimise the largest possible regret and also to maximise the minimum 'benefit'. The game theory approach therefore can be made more realistic if, instead of considering only one all-embracing criterion, several criteria are considered together and a compromise is sought among them, leading to the idea of 'compromise games'.

To explain the 'compromise games' ideas, we return to Hazell's data and treat Table 7.1 as the pay-off matrix of a game. The rows of this matrix represent the farmer's 'strategies', that is, the crops he would grow. Each column of this table corresponds to a year in the time series and a year or the time period is the state of nature resulting from the total sum of conditions that affect agriculture. We can now interpret gross margins as the 'pay-off' entries in the matrix. Thus for instance, the pay-off of $292 represents the gross margin that the farmer will achieve if he follows the strategy of growing carrots when facing the state of nature as represented by the first year of the time series. From these pay-offs the regret and 'benefit' matrices of Tables 7.4 and 7.5 are derived. These games have been reduced to their equivalent LP models by using McInerney's (1967) method as given in Table 7.6.

Table 7.4 Savage's matrix for Hazell's example

Years/Crops	t_1	t_2	t_3	t_4	t_5	t_6
Carrots (x_1)	287	460	534	677	0	591
Celery (x_2)	707	79	0	380	244	0
Cucumbers (x_3)	159	452	282	675	104	691
Peppers (x_4)	0	0	269	0	421	281

Table 7.5 Agrawal-Heady matrix for Hazell's example

Years/Crops	t_1	t_2	t_3	t_4	t_5	t_6
Carrots (x_1)	420	0	0	0	421	100
Celery (x_2)	0	381	534	297	177	691
Cucumbers (x_3)	548	8	252	2	317	0
Peppers (x_4)	707	460	265	677	0	410

Table 7.6 Linear programming models for the three different criteria for dealing with uncertainty

Linear programming model for the Wald criterion

   Maximise V

subject to

$$292x_1 - 128x_2 + 420x_3 + 579x_4 - V \geq 0$$
$$179x_1 + 560x_2 + 187x_3 + 639x_4 - V \geq 0$$
$$114x_1 + 648x_2 + 366x_3 + 379x_4 - V \geq 0$$
$$247x_1 + 544x_2 + 249x_3 + 924x_4 - V \geq 0$$
$$426x_1 + 182x_2 + 322x_3 + 5x_4 - V \geq 0$$
$$259x_1 + 850x_2 + 159x_3 + 569x_4 - V \geq 0$$

   $x \in E'$ [Technical restraints (1), (2) and (3) from Table 7.2]

Linear programming model for the Savage criterion

   Minimise R

subject to

$$287x_1 + 707x_2 + 159x_3 - R \leq 0$$
$$460x_1 + 79x_2 + 452x_3 - R \leq 0$$
$$534x_1 + 282x_3 + 269x_4 - R \leq 0$$
$$697x_1 + 380x_2 + 675x_3 - R \leq 0$$
$$244x_2 + 104x_3 + 421x_4 - R \leq 0$$
$$591x_1 + 691x_3 + 281x_4 - R \leq 0$$

   $x \in E'$ [Technical restraints (1), (2) and (3) from Table 7.2] *

Linear programming model for the Agrawal–Heady criterion

   Maximise B

subject to

$$420x_1 + 548x_3 + 707x_4 - B \geq 0$$
$$381x_2 + 8x_3 + 460x_4 - B \geq 0$$
$$534x_2 + 252x_3 + 265x_4 - B \geq 0$$
$$297x_2 + 2x_3 + 677x_4 - B \geq 0$$
$$421x_1 + 177x_2 + 317x_3 - B \geq 0$$
$$100x_1 + 691x_2 + 410x_4 - B \geq 0$$

   $x \in E'$ [Technical restraints (1), (2) and (3) from Table 7.2]

*In this set of restraints, the land restraint is an equality (i.e. $x_1 + x_2 + x_3 + x_4 = 200$), as the use of an inequality of the type ≤ would allow the objective function of minimum regret to be optimised at $R = 0$, without using any land at all.

By solving the three LP models individually and by computing the value of each criterion for all three optimal solutions, the pay-off matrix as given in Table 7.7 is obtained. An additional row has been included here in the matrix to represent the results obtained when only the expected (average) gross margin is optimised.

The degree of conflict that exists between the four criteria included in our example can be seen easily. This degree of conflict is specially strong for the Wald and Agrawal–Heady criteria. In fact, for these criteria the outcomes move from $60,456 to only $37,558 of maximum minimum gross margin and from $43,845 to $24,651 of maximum minimum 'benefit'.

The main-diagonal elements of the pay-off matrix establish an ideal point, which in our problem consists of achieving a gross margin of at least $60,456, with a largest possible regret of only $79,760, a minimum 'benefit' of at least $43,845 and an expected gross margin of $77,996. As the ideal point is infeasible we must look for a compromise between the four criteria considered through a compromise game. Before that, however, it is necessary to examine some basic ideas of a relatively new operational research technique known as games with multiple goals.

Games with multiple goals and goal programming

To illustrate the main points of the games with multiple goals let us take a simple game against nature with the following pay-off and 'regret' matrices:

$$\begin{bmatrix} 200 & 400 \\ 300 & 100 \end{bmatrix} \quad \begin{bmatrix} 100 & 0 \\ 0 & 300 \end{bmatrix}$$

Assume that the DM wants to find the 'mixed strategy', which would allow him to achieve a minimum outcome of 350 and a maximum regret of 50; and, achieving the first target is twice as important as the achievement of the second one. Cook (1976) is credited as being the first to introduce such games with more than one pay-off matrix. He titled them 'games with multiple goals'.

A straightforward method to solve this game is to reduce it to a GP model. To do that, if we take the above pay-off matrix, then the DM will be interested in a pure or mixed strategy which allows him to meet his targets, satisfying the following conditions:

$$200x_1 + 300x_2 \geq 350$$
$$400x_1 + 100x_2 \geq 350 \tag{7.4}$$

where x_1 and x_2 represent the probabilities or frequencies with which the player uses strategies 1 and 2 respectively.

Table 7.7 A pay-off matrix for the four criteria considered

	Criteria					Cropping patterns			
	Wald ($)	Savage ($)	Agrawal-Heady	Gross margin expectation ($)	Carrots x_1 (acres)	Celery x_2 (acres)	Cucumbers x_3 (acres)	Peppers x_4 (acres)	
Wald	60,456	109,472	24,651	65,309	86.32	30.69	55.78	27.21	
Savage	47,785	79,760	33,620	74,968	100	26.47	–	73.53	
Agrawal-Heady	44,845	82,097	43,845	75,754	64.17	30.91	35.83	69.09	
Gross margin expectation	37,558	89,486	36,559	77,996	–	27.45	100	72.55	

The fulfilment of inequalities (7.4) guarantees no under-achievements with respect to the first target. By adding the positive and negative deviational variables to the above inequalities of GP, and we obtain the following equalities:

$$200x_1 + 300x_2 + n_1 - p_1 = 350$$
$$400x_1 + 100x_2 + n_2 - p_2 = 350 \tag{7.5}$$

As the DM is interested in minimising under-achievement with respect to the target 350, n_1 and n_2 are minimised.

Similarly, from the 'regret' matrix we obtain the following goal constraints:

$$100x_1 + n_3 - p_3 = 50$$
$$300x_2 + n_4 - p_4 = 50 \tag{7.6}$$

In this case as the DM is interested in minimising over-achievements with respect to the regret target of 50; therefore, p_3 and p_4 are minimised.

If nature presents its first state, then the total deviation for the DM for the targets of 350 and 50 are given by the weighted sum of the negative n_1 and the positive p_3 deviational variables; thus, the total weighted deviation when nature presents its first state, is:

$$2n_1 + p_3 \tag{7.7}$$

Similarly, the total weighted deviation under the second state of nature is:

$$2n_2 + p_4 \tag{7.8}$$

Supposing that the DM is interested in finding the strategy that minimises the maximum weighted deviation. If V represents the maximum deviation then the optimum strategy is found by minimising V subject to equalities (7.5) and (7.6) and the constraints where total weighted deviations (7.7) and (7.8) are less than or equal to V. The structure of this model is:

Minimise V (7.9)

subject to

$$2n_1 + p_3 \leq V$$
$$2n_2 + p_4 \leq V$$
$$200x_1 + 300x_2 + n_1 - p_1 = 350$$
$$400x_1 + 100x_2 + n_2 - p_2 = 350$$
$$100x_1 + n_3 - p_3 = 50$$
$$300x_2 + n_4 - p_4 = 50$$
$$x_1 + x_2 = 1$$

This is a non-pre-emptive GP model with a MINMAX objective function (see Chapter 3), which corresponds to the use of the L_∞ metric ($p = \infty$). The DM can be interested in other objectives such as the minimisation of the sum of weighted deviations which correspond to the use of L_1 metric ($p = 1$). In that case the optimum strategy for the DM will be obtained by solving the following WGP model:

Minimise $2(n_1 + n_2) + (p_3 + p_4)$ (7.10)

subject to

constraints (7.5) and (7.6)

$$x_1 + x_2 = 1$$

If we set infeasible bounds on aspiration levels that cannot be achieved (e.g. in our example 400 and 0 respectively) then the game with multiple goals turns into a compromise programming (CP) model. This link between GP and CP models allows us to turn a game with multiple goals into a compromise game for solving decision problems discussed earlier in this chapter.

Compromise games

An extension of the games with multiple goals is presented here and we refer to this extension as compromise games and develop the method by continuing to work with Hazell's example farm.

A compromise between the four criteria as discussed above is established by treating the ideal value (i.e. the elements of the main diagonal of the matrix shown in Table 7.7) of each criterion as the target. The corresponding GP model then coincides with a CP structure. Using the same reasoning as the games with multiple goals, and working with relative rather than absolute deviations because of the different types of the four criteria, it is straightforward to build the LP model given in Table 7.8 which allows us to obtain the best-compromise solution, when the L_∞ metric is used (i.e. when only the largest deviation counts). Obviously the weights attached to the deviational variables reflect the importance given by the DM to the difference between the actual achievement for each criterion and its ideal value.

The solution of the above LP problem for a certain set of weights represents a compromise strategy, which is the bound of the compromise set corresponding to the metric L_∞. To obtain the optimum strategy for the other bound of the compromise set we only need to undertake the optimisation process according to the L_1 metric. To do that, it is only necessary to eliminate the *game constraints* of the LP matrix of Table 7.8 and to introduce the objective function corresponding to the L_1 metric. On making these alterations, we obtain the following LP model:

$$\text{Min } L_1 = w_1 \, \frac{n_1 + n_2 + n_3 + n_4 + n_5 + n_6}{60{,}456 - 37{,}558} + w_2 \, \frac{p_7 + p_8 + p_9 + p_{10} + p_{11} + p_{12}}{109{,}472 - 79{,}760}$$

$$(7.11)$$

$$+ w_3 \, \frac{n_{13} + n_{14} + n_{15} + n_{16} + n_{17} + n_{18}}{43{,}845 - 24{,}651} + w_4 \, \frac{6n_{19}}{77{,}996 - 65{,}309}$$

subject to technical constraints and game-goal constraints from Table 7.8.

For a given set of weights representing the preference of the DM for the four criteria, we can obtain the compromise strategies as a subset of the set of efficient strategies which are closest to the ideal.

Tables 7.9 and 7.10 show the solutions generated for the metrices L_1 and L_∞ and for different sets of weights. The compromise strategies are presented in the criteria and in the decision variable space. It should be pointed out that the compromise strategies that have been generated look very promising as in all but one case the performance of any criterion is not below 80% of its ideal value. Only for the Wald criterion some compromise solutions are around 70% of their ideal.

Table 7.8 Linear programming matrix for the game with multiple pay-offs (L_∞ metric)

Objective function: Minimise $L_\infty = d$

subject to

Game constraints:

$$w_1 \frac{n_1}{60{,}456 - 37{,}558} + w_2 \frac{p_7}{109{,}472 - 79{,}760} + w_3 \frac{n_{13}}{43{,}845 - 24{,}651} + w_4 \frac{n_{19}}{77{,}996 - 65{,}309} \leq d$$

$$w_1 \frac{n_2}{60{,}456 - 37{,}558} + w_2 \frac{p_8}{109{,}472 - 79{,}760} + w_3 \frac{n_{14}}{43{,}845 - 24{,}651} + w_4 \frac{n_{19}}{77{,}996 - 65{,}309} \leq d$$

$$w_1 \frac{n_3}{60{,}456 - 37{,}558} + w_2 \frac{p_9}{109{,}472 - 79{,}760} + w_3 \frac{n_{15}}{43{,}845 - 24{,}651} + w_4 \frac{n_{19}}{77{,}996 - 65{,}309} \leq d$$

$$w_1 \frac{n_4}{60{,}456 - 37{,}558} + w_2 \frac{p_{10}}{109{,}472 - 79{,}760} + w_3 \frac{n_{16}}{43{,}845 - 24{,}651} + w_4 \frac{n_{19}}{77{,}996 - 65{,}309} \leq d$$

$$w_1 \frac{n_5}{60{,}456 - 37{,}558} + w_2 \frac{p_{11}}{109{,}472 - 79{,}760} + w_3 \frac{n_{17}}{43{,}845 - 24{,}651} + w_4 \frac{n_{19}}{77{,}996 - 65{,}309} \leq d$$

$$w_1 \frac{n_6}{60{,}456 - 37{,}558} + w_2 \frac{p_{12}}{109{,}472 - 79{,}760} + w_3 \frac{n_{18}}{43{,}845 - 24{,}651} + w_4 \frac{n_{19}}{77{,}996 - 65{,}309} \leq d$$

Technical constraints:

$$x_1 + x_2 + x_3 + x_4 = 200$$
$$25x_1 + 36x_2 + 27x_3 + 84x_4 \leq 10{,}000$$
$$x_1 + x_2 + x_3 + x_4 \leq 0$$

Game goal constraints:
Wald's criterion

$$292x_1 + 128x_2 + 420x_3 + 579x_4 + n_1 - p_1 = 60{,}456$$
$$179x_1 + 560x_2 + 187x_3 + 639x_4 + n_2 - p_2 = 60{,}456$$
$$114x_1 + 648x_2 + 366x_3 + 379x_4 + n_3 - p_3 = 60{,}456$$
$$247x_1 + 544x_2 + 249x_3 + 924x_4 + n_4 - p_4 = 60{,}456$$
$$426x_1 + 182x_2 + 322x_3 + 5x_4 + n_5 - p_5 = 60{,}456$$
$$259x_1 + 850x_2 + 159x_3 + 569x_4 + n_6 - p_6 = 60{,}456$$

Savage's criterion

$$287x_1 + 707x_2 + 159x_3 + n_7 - p_7 = 79{,}760$$
$$460x_1 + 79x_2 + 452x_3 + n_8 - p_8 = 79{,}760$$
$$534x_1 + 282x_3 + 269x_4 + n_9 - p_9 = 79{,}760$$
$$697x_1 + 380x_2 + 675x_3 + n_{10} - p_{10} = 79{,}760$$
$$244x_2 + 104x_3 + 421x_4 + n_{11} - p_{11} = 79{,}760$$
$$591x_1 + 691x_3 + 281x_4 + n_{12} - p_{12} = 79{,}760$$

Agrawal–Heady criterion

$$420x_1 + 548x_3 + 707x_4 + n_{13} - p_{13} = 43{,}845$$
$$381x_2 + 8x_3 + 460x_4 + n_{14} - p_{14} = 43{,}845$$
$$534x_2 + 252x_3 + 265x_4 + n_{15} - p_{15} = 43{,}845$$
$$297x_2 + 2x_3 + 677x_4 + n_{16} - p_{16} = 43{,}845$$
$$421x_1 + 177x_2 + 317x_3 + n_{17} - p_{17} = 43{,}845$$
$$100x_1 + 691x_2 + 410x_4 + n_{18} - p_{18} = 43{,}845$$

Gross margin expectations

$$253x_1 + 443x_2 + 284x_3 + 516x_4 + n_{19} - p_{19} = 77{,}996$$

Table 7.9 Sensitivity analysis of the compromise game with multiple pay-offs (L_∞ metric)

Set of weights used for computer runs 1 to 4	Criteria				Cropping patterns			
	Wald ($)	Savage ($)	Agrawal-Heady ($)	Gross margin expectation ($)	Carrots x_1 (acres)	Celery x_2 (acres)	Cucumbers x_3 (acres)	Peppers x_4 (acres)
1. $w_1 = w_2 = w_3 = w_4$	51,763	91,032	40,715	72,873	86.32	51.81	16.56	45.31
2. $w_1 = 2; w_2 = w_3 = w_4 = 1$	54,854	93,911	36,461	70,895	104.75	42.05	7.29	45.92
3. $w_1 = 3; w_2 = w_3 = w_4 = 1$	55,620	95,580	35,754	70,214	104.57	40.20	10.99	44.24
4. $w_1 = w_2 = w_3 = 1; w_4 = 2$	45,334	94,624	40,776	74,789	–	71.38	100	28.62

Table 7.10 Sensitivity analysis of the compromise game with multiple pay-offs (L_1 metric)

Set of weights used for computer runs 1 to 4	Criteria				Cropping patterns			
	Wald ($)	Savage ($)	Agrawal-Heady ($)	Gross margin expectation ($)	Carrots x_1 (acres)	Celery x_2 (acres)	Cucumbers x_3 (acres)	Peppers x_4 (acres)
1. $w_1 = w_2 = w_3 = w_4$	42,817	84,323	42,191	76,355	58.04	25.17	41.95	74.82
2. $w_1 = 7; w_2 = w_3 = w_4 = 1$	49,527	84,179	37,657	74,209	96.91	38.30	3.49	61.70
3. $w_1 = w_4 = 2; w_2 = w_3 = 1$	44,204	82,959	43,204	75,990	61.79	28.69	38.21	71.31
4. $w_1 = w_4 = 3; w_2 = w_3 = 1$	44,845	82,097	43,845	75,754	64.17	30.91	35.83	69.09

It is also interesting to note the strong competition between Wald and gross margin expectation criteria and the degree of complementarity between this and the Agrawal–Heady criterion. Finally, it is interesting to note that with L_1 metric the compromises obtained are better in terms of Savage, Agrawal–Heady and gross margin expectations.

Suggestions for further reading

The theory of games is a fundamental contribution to the theory of economic behaviour. It appeared on the scene in 1944 when von Neumann and Morgenstern published their monumental work, *Theory of Games and Economic Behaviour*. The application of the theory of games to the analysis of agricultural decision-making under uncertainty was initiated by McInerney (1967, 1969). McInerney's contributions can be considered as a 'rejoinder' to the pessimistic view promoted by Dillon (1962) on the possibilities of the practical applications of the theory of games to agricultural decision-making. In 1968 Agrawal and Heady presented the 'benefit' criterion as a compromise between the pessimism of Wald criterion and the optimism of Savage criterion. Hazell (1970) introduced the idea of parametric games, setting the gross margin expectation as a restraint, which is varied parametrically. Kawaguchi and Maruyama (1972) introduced a game with two criteria (Wald and Savage) within an agricultural planning context without using the theoretical framework of games with multiple goals.

Romero and Rehman (1985, pp. 181–185) have shown how the traditional agricultural planning models under risk and uncertainty conditions can be incorporated within the MCDM framework. An interesting application of compromise risk programming in farm management is Hope and Lingard (1992). In this context the approach mean-partial absolute deviation that connects safety-first models with compromise risk programming should be cited (Berbel 1988). A good state-of-the-art survey of classic risk programming techniques in agriculture can be found in the book edited by Hardaker et al. (1997).

The first formulation of a two-person zero-sum game with multiple goals by Cook (1976) has been simplified considerably by Hannan (1982); although both authors do not seem to realise the crucial link between such games and Goal Programming. Zeleny (1976, 1982 pp. 358–360) has handled a similar problem within a multiobjective programming approach. The Compromise Games against nature proposed in the above chapter could be considered a natural extension of these works.

Colson and Zeleny (1979, 1980) and Zeleny (1982 Chap.11) have proposed a new approach to risk programming, termed as Prospect Ranking Vector, where the measure of risk is multidimensional. This approach is made operational using multiobjective programming techniques. A linkage between this approach and risk programming in agriculture can be seen in Romero (2000).

Notes

1   The first part of this chapter is based on a paper by Romero et al. (1988).

2   As has been shown in the last section of Chapter 5, the use of the L_1 metric in CP implies the use of a linear and additive utility function.

Part three
Case studies

A number of carefully selected case studies in the application of the MCDM paradigm to agricultural decision-making are presented here. The common feature of these studies is that they are sufficiently general and detailed to serve as a basis for appreciating the application of the various MCDM techniques that have been explained in part two to specific decision situations.

Chapter eight

A compromise programming model for the agrarian reform programme in Andalusia, Spain

One of the main objectives of the 1984 Agrarian Reform Law (ARL) for Andalusia was to mitigate the high rate of unemployment currently experienced by the rural areas in this part of Spain. Rural unemployment had reached such alarming levels that it had led to a general peasant unrest resulting in the illegal takeover by farm workers of large agricultural holdings. The ARL empowers the Andalusian Institute of Agrarian Reform (IARA) to implement, among other measures, the expropriation of rural holdings under certain conditions of low productivity indices (level of production, employment, etc.). Once the holdings had been expropriated they would be redistributed among the workers organised in co-operatives. The IARA will then recommend the farm plans to be established in these co-operatives.

Given this situation, the IARA must pay special attention to the employment level in the co-operative when choosing the optimum cropping pattern. However, this objective can be interpreted either as the level of daily wages paid throughout the year or as the number of permanent workers employed in the co-operative. Unfortunately, these two objectives are conflicting ones for crops cultivated in the irrigated lands of Andalusia. Providing permanent jobs throughout the year for a maximum number of workers can only be achieved for relatively low employment. In contrast, high employment (daily wages) can only be achieved with high seasonal labour. Moreover, the maximisation of business profitability in the co-operative, presumably the main objective of its members, and the maximisation of stable employment are also in conflict. However, business profitability and employment seem to be almost complementary objectives (see Table 8.2) and, therefore, these objectives clash with each other; the social objective of stable employment is also in conflict with the private objective of business profitability.

The purpose of this chapter (based on Romero et al. 1987) is to show how the above conflict could have been resolved by using MOP techniques to find a compromise between employment, seasonal labour use and business profitability. As a by-product of this analysis the opportunity costs, in terms of business profitability and employment, of a stable employment policy are also determined.

Background

We take the case of a co-operative on an irrigated arable farm of 100 ha under the agrarian reform programme of Andalusia. Table 8.1 shows the LP matrix for this farm planning problem. Most of the constraints in the matrix are self-explanatory although some require further explanation. Constraints (14) – (17) for seasonal labour represent the deviations between labour utilisation for each crop in each of the four quarters and the average labour utilisation for each crop. Thus, the first coefficient of row (14) is -59.49 since cotton requires 4.14 hours/ha in the first quarter while the average labour utilisation for this crop is 63.63 hours/ha per quarter. The deviational variables x_{14} to x_{21} measure the under- and over-achievements with respect to a null deviation in each of the quarters considered. The minimisation of the sum of the deviational variables implies the minimisation of the mean absolute deviation (see Chapter 7 and the MOTAD method); hence, the minimisation of the objective function Z_1 given at the bottom of Table 8.1 implies the minimisation of seasonal labour.

Constraints (18) – (21) guarantee positive cash flows in every quarter. The possibility of transferring the cash surplus from one quarter to the next is included in all but the last quarter where only 25% of the possible surplus can be transferred.

The three objectives that have been considered are minimisation of seasonal labour use, maximisation of employment and maximisation of gross margin as expressed in the last three rows of Table 8.1.

A device frequently used within the MOP approach is the pay-off matrix (see Chapter 4). To obtain the elements of this matrix, the *ordinary* linear programming model is solved as many times as there are objectives under consideration in the problem. For each solution, the optimal value of the objective being optimised is accompanied by the values of the remaining objectives at that solution. The pay-off matrix is very useful to illustrate the degree of conflict between the objectives under consideration. Table 8.2 shows the pay-off matrix for the three objectives. The elements of that matrix are easy to understand. For example, the elements in the first row mean that the minimum seasonal labour solution (15.97 hours/ha) corresponds to an employment of 156.18 hours/ha and a gross margin of 82,321 pesetas/ha.

Table 8.1 The multicriteria linear programming model for an irrigated arable farm in Seville (Andalusia)*

Real activities (ha)				Seasonal labour deviational variables (hours)					Working capital transfer activities (thousand pesetas)						
Cotton	Wheat	Lettuces + corn	Peach trees												
x_1	x_2 ...	x_{12}	x_{13}	x_{14}	x_{15} ...	x_{20}	x_{21}	x_{22} ...	x_{24}	x_{25}					
1											≤	30	Cotton maximum (ha)	[1]
			1								≤	15	Peach trees maximum (ha)	[8]
10	2	58	4.5								≤	750	Tractor hours – period 1	[9]
2	9	52	6								≤	750	Tractor hours – period IV	[12]
1	1	1	1								=	100	Crop land (ha)	[13]
-59.49	-4.57	124.78	-35.94	1	-1						=	0	Seasonal labour (hours) – period I	[14]
21.29	-3.24	96.79	-235.05			1	-1				=	0	Seasonal labour (hours) – period IV	[17]
-2.95	-3.21	86.04	-87.46					-1	0.25		≥	0	Working capital (thousand pta/ha) – Period 1	[18]
252.04	-3.27	140.34	-30.83						1	-1	≥	0	Working capital (thousand pta/ha)-Period IV	[21]
				1	...	1	1							Z_1 Minimise seasonal labour (hours)
254.50	26.25	904.40	1356											Z_2 Maximise employment (hours)
107.13	53.78	275.52	332.88											Z_3 Maximise gross margin (thousand pta)

*A complete description of the matrix can be seen in Romero et al. (1987, p. 81)

Table 8.2 Payoff matrix for the three objectives

	Seasonal labour* (hours/ha)	Employment (hours/ha)	Gross margin (pesetas/ha)
Seasonal labour	*15.97*	156.18	82,321
Employment	225.28	*451.9*	172,107
Gross margin	229.9	421.13	*174,116*

*Seasonal labour is measured as the mean absolute deviation for the four quarters.

From Table 8.2 it is easy to see that the seasonal labour objective clearly conflicts with employment and gross margin, although the latter two are almost complementary. In fact, the maximum gross margin (174,116 pesetas/ha) is nearly identical to the gross margin when the employment objective is optimised (172,107 pesetas/ha). Similarly, employment levels are about the same whether employment is optimised (451.90 hours/ha) or gross margin is optimised (421.73 hours/ha).

The elements in the main diagonal of the pay-off matrix establish the ideal point; this is the point where all the objectives achieve their optimum value. In this problem the ideal point is 15.97 hours/ha for seasonal labour; 451.90 hours/ha for employment and 174,116 pesetas/ha for gross margin. However, the ideal point is infeasible, since the objectives are in conflict; therefore, it is only possible to choose the point of minimum seasonal labour, or the maximum level of employment, or the maximum level of gross margin or a compromise between these points.

Trade-off curves for seasonal labour, employment and gross margin

The points of minimum seasonal labour and maximum gross margin are considered to be the bounds of a transformation curve which measures the relationship between both objectives. Similarly, the points of minimum seasonal labour and maximum employment are considered to be the bounds of another transformation curve that measures the relationship between these objectives. Generating this curve indicates the trade-offs between the objectives being considered, which can be characterised as the opportunity cost of seasonal labour in terms of gross margin and in terms of employment, and vice versa. computational terms, deriving the transformation curve is equivalent to the generation of the set of efficient or Pareto optimal solutions.

We have chosen a variant of the weighting method, the noninferior set estimation (NISE), because it permits the exact generation of the efficient set when only two objectives are involved, as explained in Chapter 4. On applying the NISE method for the bi-criteria LP problems, the trade-off curves or efficient sets between labour seasonality and employment, (Figure 8.1), and labour seasonality and gross margin (Figure 8.2), are obtained. The co-ordinates of these extreme points and the values of the decision variables (cropping patterns) are shown in Tables 8.3 and 8.4.

The actual values of the trade-offs (i.e. the opportunity costs) between seasonal labour-employment and between seasonal labour-gross margin are represented by the slopes of the straight lines connecting the extreme efficient points shown in Figures 8.1 and 8.2. For example, the slope of segment AB in Figure 8.1 indicates that in this part of the trade-off curve each hour/ha increase in seasonal labour increases employment by 5.81 hours/ha. Thus, the opportunity cost of one hour/ha of more even labour utilisation can be measured as a sacrifice of 5.81 hours/ha of employment. Similarly, the slope of segment FG (Figure 8.2) indicates that in this part of the curve the opportunity cost of one hour/ha of more even labour utilisation has a gross margin of 1,590 pesetas/ha.

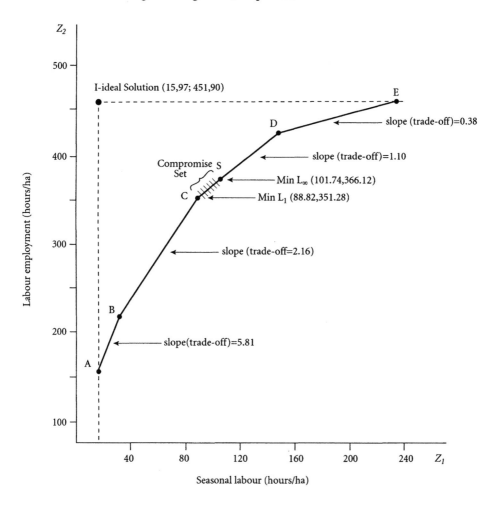

Figure 8.1 The trade-off curve for seasonal labour and employment

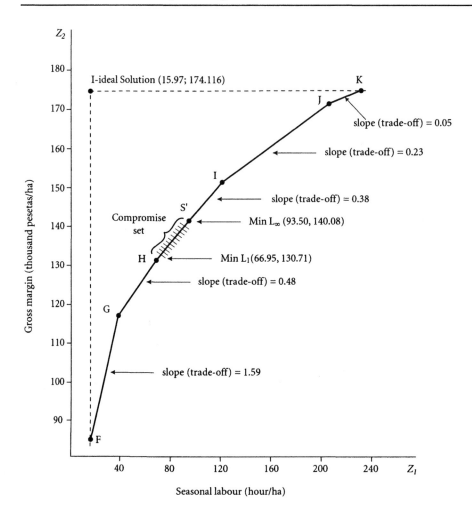

Figure 8.2 The trade-off curve for seasonal labour and gross margin

The IARA should choose the optimum farm plan from the sets of efficient solutions represented by the trade-off curves. But, which efficient farm plan would it be? The answer depends on the preferences that the IARA attach to each objective; that is, it depends on the subjective values of the trade-offs between the objectives. For instance, suppose the IARA chooses the farm plan given by point G instead of the plan given by point F (see Figure 8.2). This choice means that for the IARA a reduction of 21.49 hours/ha of seasonal labour does not compensate a decrease of 34,110 pesetas/ha of gross margin.

Compromise sets

Once the trade-off curves or efficient sets have been defined, the next task is to establish the optimum efficient point or at least to reduce the size of the efficient sets. It is undertaken by using compromise programming (CP) as explained in Chapter 5.

Recall that for CP modelling the objectives measured in different units (pesetas of gross margin, hours of employment and hours of labour seasonality), relative deviations rather than absolute ones must be used to establish the degree of closeness between each objective and its ideal point. The ideal values for the three objectives are represented in the main diagonal of the pay-off matrix shown in Table 8.2. The anti-ideals or nadir points correspond to the minimum elements of columns 2 and 3 and the maximum element of column 1 of the pay-off matrix shown in Table 8.2 (i.e. 156.18 hours/ha of employment, 82,321 pts/ha of gross margin and 229.90 hours/ha of labour seasonality).

Applying the above figures to the LP models presented in Chapter 5, the best compromise solutions for the metrics L_1 and L_∞ can be obtained for the situation described in Table 8.1. Thus, for the metric L_1 (that is, for $p = 1$), the best-compromise solution is found by solving the following LP problems:

Labour seasonality–employment

$$\text{Min } L_1 = \frac{w_1 \left[Z_1(\underline{x}) - 15.97 \right]}{229.90 - 15.97} + \frac{w_2 \left[451.90 - Z_2(\underline{x}) \right]}{451.90 - 156.18}$$

subject to

$$\underline{x} \, \varepsilon \, \underline{F} \, [\text{technical constraint from Table 8.1}]$$

Labour seasonality–gross margin

$$\text{Min } L_1 = \frac{w_1' \left[Z_1(\underline{x}) - 15.97 \right]}{229.90 - 15.97} + \frac{w_2' \left[174.116 - Z_3(\underline{x}) \right]}{174.116 - 82.230}$$

subject to

$$\underline{x} \in \underline{F} \, [\text{technical constraint from Table 8.1}]$$

Table 8.3 Feasible efficient extreme points and cropping patterns (labour seasonality–employment)

Extreme points potatoes	Objective functions		Decision variables													
	Z_1 (labour seasonality) (hours/ha)	Z_2 (employment) (hours/ha)	x_1 cotton (ha)	x_2 wheat (ha)	x_3 corn (ha)	x_4 soybean (ha)	x_5 potatoes (ha)	x_6 sugar beet (ha)	x_7 soybean + wheat (ha)	x_8 potatoes+ soybean (ha) + corn (ha)	x_9 soybean (ha) + corn (ha)	x_{10} sorghum (ha)	x_{11} lettuces (ha)	x_{12} lettuces + corn (ha)	x_{13} peach trees (ha)
A	15.97	156.18	16.43	25	12.45	25	2.10	4.05	–	–	–	8.56	6.41	–	–
B	26.32	216.31	30	24.97	1.59	25	–	1.09	–	–	–	6.19	6.94	–	4.22
$L_1 \rightarrow$C	88.82	351.28	30	1.83	4.57	25	–	20	–	–	–	4.04	4.86	–	9.7
D	151.03	419.68	30	–	3.2	22.4	–	20	–	–	–	4.7	4.7	–	15
E	235.28	451.9	30	–	–	–	–	20	–	14.13	0.87	19.81	–	0.19	15
$L_\infty \rightarrow$S	101.74	366.12	30	1.09	4.39	25	–	20	–	–	–	3.82	4.86	–	10.84

Table 8.4 Feasible efficient extreme points and cropping patterns

Extreme points potatoes	Objective functions		Decision variables													
	Z_1 (labour seasonality) (hours/ha)	Z_2 (employment) (hours/ha)	x_1 cotton (ha)	x_2 wheat (ha)	x_3 corn (ha)	x_4 soybean (ha)	x_5 potatoes (ha)	x_6 sugar beet (ha)	x_7 soybean + wheat (ha)	x_8 potatoes+ soybean (ha) + corn (ha)	x_9 soybean (ha) + corn (ha)	x_{10} sorghum (ha)	x_{11} lettuces (ha)	x_{12} lettuces + corn (ha)	x_{13} peach trees (ha)
F	15.97	82.320	16.43	25	12.45	25	2.1	4.05	–	–	–	8.56	6.41	–	–
G	37.46	116.430	27.72	11.57	–	25	15	1.92	–	–	–	13.66	4.64	–	0.49
$L_1 \rightarrow$H	66.95	130.709	30	–	11.49	–	15	8.37	25	–	–	9.72	0.42	–	–
I	121.23	151.088	30	3.63	13.66	10	–	20	–	15	–	1.08	–	–	6.63
J	216.71	173.424	30	–	12.08	8.91	–	19.01	–	15	–	–	–	–	15
K	229.90	174.116	30	–	8.81	–	–	12.83	10	15	–	8.36	–	–	15
$L_\infty \rightarrow$S'	93.50	140.080	30	–	15.8	6.56	11.11	14.78	14.55	3.89	–	–	–	–	3.31

The optimum solutions of the above LP problems for $w_1 = w_2$ and $w_1' = w_2'$ (that is, when the objectives are equally important) is given by the points C and H for the first and for the second transformation curve respectively; therefore, the points C and H are the best-compromise solutions and this means that C and H are the efficient points closest to the ideal points when the metric L_1 is used.

For the metric L_∞ (that is, for $p = \infty$), the best-compromise solution is found by solving the following problems:

Labour seasonality–employment

Minimise $L_\infty = d$

subject to

$$\frac{w_1[Z_1(\underline{x}) - 15.97]}{229.90 - 15.97} \le d$$

$$\frac{w_2[451.90 - Z_2(\underline{x})]}{451.90 - 156.18} \le d$$

$\underline{x} \in F$ [technical constraint from Table 8.1]

Labour seasonality–gross margin

Minimise d'

subject to

$$\frac{w_1'[Z_1(\underline{x}) - 15.97]}{229.90 - 15.97} \le d'$$

$$\frac{w_2'[174.116 - Z3(x)]}{174.116 - 82.320} \le d'$$

$\underline{x} \in F$ [technical constraint from Table 8.1]

where d and d' are the largest deviations. The optimum solution of the above LP problems, assuming again $w_1 = w_2$ and $w_1' = w_2'$, is given by points S and S' for the first and for the second transformation curves respectively. Figures 8.1 and 8.2 and the last rows of Tables 8.3 and 8.4 show the values of z_1, z_2 and z_3 and the farm plan corresponding to points S and S'.

In Chapter 5 it was shown that metrics L_1 and L_∞ define a subset of the efficient set called compromise set. The other best-compromise solutions (for $1 < p < \infty$) fall between the solutions corresponding to metrics L_1 and L_∞; therefore, the segments CS and HS' represent the compromise sets. Obviously, if the weights w_j and w_j' attached to the discrepancies between each objective and its ideal value are different with respect to the values considered in this case, the structure of the compromise sets can be modified. A sensitivity analysis with the w and w' weights can furnish the decision maker with worthwhile data related to the stability of the solution and to the range within which the compromise sets can be defined.

An approximation of the efficient set in a three-dimensional objectives space

We now examine the trade-offs and compromise sets when the three objectives are considered simultaneously. For this, the efficient set for the three objectives must be established and to do that we have used the constraint method. In this case, seasonal labour was chosen as the objective function while the two other objectives were treated as the parametric restraints giving the following LP model:

Minimise $Z_1(\underline{x})$

subject to

$Z_2(\underline{x}) \geq b_1$
$Z_3(\underline{x}) \geq b_2$
$\underline{x} \in F$

The efficient set is generated by parametric variations of b_1 and b_2. The initial lower bound for b_1 was set at 160 hours/ha of employment and the final lower bound at 450 hours/ha. The increments for b_1 were 40 hours/ha. The initial lower bound for b_2 was set at 85,000 pesetas/ha of gross margin and the final lower bound at 170,000 pesetas/ha. The increments for b_2 were 10,000 pesetas/ha. Solving the corresponding LP problems resulted in more than 50 efficient points.

Presenting a decision maker with more than 50 possible cropping patterns could be an 'overload' of information that he may find difficult to assimilate. In order to alleviate this possible problem, a feature of MOP applications, we have used a filtering technique proposed by Steuer and Harris (1980) to reduce the size of the efficient set. This filtering technique discards efficient solutions that are not sufficiently different from other efficient solutions already calculated and retained by the filter. The filtering relationship used is:

$$\left[\sum_{j=1}^{k} (A_j |z_j^t - z_j^h|)^p \right]^{1/p} < d$$

where k is the number of objectives; A_j are the graduation weights used to scalarise the different objectives; Z_j^i is an efficient solution in the objective space and its dissimilarity with respect to solution Z_j^h is tested; d is the test-distance parameter used and p is the metric parameter. In our case the filtering process was implemented for the $p = 1$ metric (that is, for L_1). The appropriate value for d is established experimentally according to the size of the desired cluster. Applying that filtering relationship to our more than 50 points, a manageable cluster of only seven efficient points was obtained. This cluster represents the seven most diverse efficient points.

Table 8.5 shows the cluster of efficient points in the objective as well as the decision space. The best-compromise solutions are given at the bottom of Table 8.5. The filter discards both of these solutions, since they are very similar to other solutions already retained by the filter. In fact, the best-compromise solution for the L_1 metric is almost identical to the efficient solution 7 and for the L_∞ metric the best-compromise solution is very similar with respect to the efficient solution 4.

A measure of the trade-offs between the three objectives can be obtained from Table 8.5. For instance, if the decision maker has not decided yet between solutions 4 and 6, he will prefer the former if he considers an increase of 80 hours/ha of employment and 40,000 pesetas/ha of gross margin more important than an increase of 92.20 hours/ha of seasonal labour. In the same way, solution 1 will be preferred to solution 2 if for the decision maker a sacrifice of 54.24 hours/ha of employment is compensated by an increase of 20,000 pesetas/ha of gross margin and by a decrease of 3.12 hours/ha of seasonal labour.

Concluding comments

We are fully aware that the empirical outcome of this exercise is constrained to the particular case analysed. It would be necessary to analyse different real situations for the irrigated lands in Andalusia to identify the basic relationships between stable employment policy in the Andalusian rural sector and the opportunity costs of this policy in terms of employment and business profitability. Anyway, from the analysis undertaken the following tentative conclusions can be deduced.

It would have been possible for the IARA to implement a stable employment policy but at a very high cost in terms of gross margin and employment. Thus, in the neighbourhood of the point of minimum seasonal labour the opportunity cost of this objective with respect to employment and gross margin was of 5.81 hours/ha and 1,590 pesetas/ha respectively.

Policies with high employment and business profitability were only compatible with high seasonal labour. Thus the compromise sets provided by CP analysis guaranteed acceptable levels of gross margin (about 130,000 pesetas/ha) and of employment (about 350 hours/ha) compatible with suitable seasonal labour. A good example of sensible compromise is point 4 (Table 8.5) which offers 120,000 pesetas/ha of gross margin and 320 hours/ha of employment with a seasonal labour of only 79.07 hours/ha.

Table 8.5 Cluster of efficient points and cropping patterns for the three objectives

Extreme points	Objective functions			Decision variables												
	Z_1 (labour seasonality) (hours/ha)	Z_2 (employment) (hours/ha)	Z_3 (gross margin) ('000 pta/ha)	x_1 cotton (ha)	x_2 wheat (ha)	x_3 corn (ha)	x_4 soybean (ha)	x_5 potatoes wheat (ha)	x_6 sugar beet soybean (ha)	x_7 soybean + corn (ha)	x_8 potatoes + (ha)	x_9 potatoes +	x_{10} sorghum	x_{11} lettuces (ha) corn (ha)	x_{12} lettuces + (ha)	x_{13} peach trees
1	42.4	185.76	120	30	15.5	10.7	16.83	15	1.12	8.17	–	–	–	2.75	–	–
2	39.28	240	100	30	18.9	1.66	25	2.06	4.59	–	–	–	6.8	6.25	–	4.75
3	66.79	240	130	30	–	5.46	11.61	15	9.3	13.39	–	–	11.74	1.99	–	1.51
4	79.07	320	120	30	2.4	2.81	25	4.9	16.06	–	–	–	6.9	4.57	–	7.36
5	134.99	400	130	30	–	2.01	25	0.91	20	–	–	–	4.3	2.28	2.18	13.32
6	171.27	400	160	30	–	9.86	11.86	–	20	–	13.14	–	2.74	–	0.65	11.75
7	211.15	440	170	30	–	11.8	9.46	–	20	–	13.61	–	–	–	0.22	14.96
L_1	216.71	440.13	173.42	30	–	12.1	8.91	–	19.01	–	15	–	–	–	–	15
L_∞	102.63	335.04	137.84	30	–	7.95	22.54	8.37	19.36	–	2.46	–	–	–	2.11	7.21

Chapter nine

Livestock ration formulation and multiple criteria decision-making techniques

One of the most successful areas of application for the conventional linear programming (LP) paradigm has been in the search for the least cost combination of foods that will meet a specified level of nutritional requirements for livestock. This procedure depends on the following fundamental assumptions:

1 there is a single objective (usually the minimisation of the cost of the diet) which is a mathematical function of the decision variables;

2 the decision variables of the problem are the amounts of the available ingredients that will constitute the diet;

3 the nutritional requirements are convertible to mathematical functions of the decision variables and they form the constraint set of the problem; and

4 the optimum diet is the one that minimises the single specified objective without any violation of the constraints imposed.

Although LP has been used widely in practice with noticeable success, the above assumptions remain its weaknesses. In formulating rations, the decision maker is interested in an economically optimal ration that achieves a compromise between several conflicting objectives such as minimisation of cost, imbalance of nutrient supplies, food bulk, and so on. A further weakness of LP is the rigidity with which given nutritional requirements have to be met. In many real-life situations these requirements do not really impose nutritional constraints that cannot be violated under any circumstances. On the contrary, a certain relaxation of these constraints imposed would not seriously affect an animal's physical and economic performance.

We believe that these methodological weaknesses of LP can be overcome substantially, and its application to diet formulation can be improved considerably, if the problem is analysed within the MCDM framework. This chapter[1] transfers the ration formulation problem from a context where the DM wishes to optimise a single objective to one where several objectives are to be optimised simultaneously. The goal programming (GP) and the multiobjective programming (MOP) techniques explained earlier are used. The problem of relaxing the rigidity of the nutritional requirements is dealt with in the next chapter.

Table 9.1 Crabtree's modified matrix

	Silage (kg DM)	Straw (kg DM)	Distillers' wet grain (kg DM)	Swedes (kg DM)	Barley (kg DM)	Dairy Compound (kg DM)	Transfer Activity		= Z minimum (£/tonne)	
	48.2	26.8	69.8	92.6	105.6	162.5				
	10.3	6.5	10.3	12.8	13	12.8	0.0	≥	213.3 Metabolisable energy – ME (MJ)	[1]
	108	32	138.6	97.2	105.3	107.2	0.0	≥	1663.3 Rumen-degradable protein – RDP (g)	[2]
	19	8	59.4	10.8	11.7	52.8	0.0	≥	620.5 Undegradable protein – UDP (g)	[3]
	3.9	2.7	1.7	3.6	0.5	11.4	0.0	≥	73.6 Calcium – Ca (g)	[4]
	3.3	0.9	3.7	3.2	4	8.5	0.0	≥	67.8 Phosphorus – P (g)	[5]
	1.6	0.7	1.4	1.2	1.3	3.5	0.0	≥	24.0 Magnesium – Mg (g)	[6]
	3.8	1.1	0.9	2.6	0.2	5.4	0.0	≥	26.9 Sodium – Na (g)	[7]
	5.7	3.2	10	3.8	4.8	19	0.0	≥	180.0 Copper – Cu (mg)	[8]
	0.09	0.04	0.2	0.04	0.04	1.5	0.0	≥	2.0 Cobalt – Co (mg)	[9]
	1.0	1.0	1.0	1.0	1.0	1.0	0.0	≤	19.0 Dry matter intake – DM (kg)	[10]
	1.0	1.0	1.0	1.0	1.0	1.0	-1.0	=	0.0 Tie line	[11]
	0.56	0.35	0.56	0.7	0.71	0.7	-0.637	≤	0.0 Metabolisability minimum q	[12]
	0.56	0.35	0.56	0.7	0.71	0.7	-0.642	≤	0.0 Metabolisability maximum q + r	[13]
	-0.8	-0.8	0.2	0.2	0.2	0.2	0.0	≤	0.0 Long roughage (%)	[14]
	0	0	0	1.0	0	0	0.0	≤	2.0 Swedes (kg)	[15]
	-0.2	-0.2	0.8	-0.2	-0.2	-0.2	0.0	≤	0.0 Distillers' grain (%)	[16]
	1	0	0	0	0	0	0.0	≤	6.5 Silage (kg)	[17]

Source: Crabtree (1982, p.30) with modifications to right-hand side parameters for dry matter and swedes and the proportionate amount of distillers' wet grain that should be included in the ration.

A livestock ration formulation example

This section adapts a diet formulation problem (Crabtree 1982), which is used as a basis for discussion in the rest of the chapter. Crabtree's example has been chosen because it is well thought out from the animal nutritional standpoint and also because it seems to have many of the problems whose solution we wish to illustrate by using the MCDM techniques.

Crabtree's LP model analyses the ration formulation of a 600 kg dairy cow with a milk yield of 30 kg/day, without any change in the liveweight of the animal. The ingredients that are available for making up the diet, along with their compositions, and nutritional requirements of the animal are given in Table 9.1. However, for our purposes we have relaxed the upper limits on total dry matter intake and the amount of swedes and distillers' wet grain in the ration somewhat, as well as the proportion of dry matter derived from silage and straw.

Most of Table 9.1 is self-explanatory but the constraint (14) means that at least 20% of the dry matter intake must be silage and straw, while the constraint (16) implies that the dry matter intake from distillers' wet grains must not exceed 20% of the total dry matter.

The least-cost diet for this problem is comprised of the following:

Silage (x_1) = 4.981 kg   Straw (x_2) = 0.0 kg

Distillers' wet grains (x_3) = 3.416 kg   Swedes (x_4) = 2.0 kg

Barley (x_5) = 2.778 kg   Dairy compound (x_6) = 5.069 kg

The diet reported in the above solution and the subsequent ones discussed in this paper are all given in units of kg dry matter per day.

Table 9.2 Nutrient requirements and ration composition using the least-cost criterion only

Ingredients	Requirements	Ration
Metabolisable energy (MJ)	213.3	213.3
Rumen-degradable protein (g)	1663.9	2045
Undegradable protein (g)	620.5	620.5
Calcium (g)	73.6	91.6
Phosphorus (g)	67.8	89.8
Magnesium (g)	24	36.5
Sodium (g)	26.9	55.1
Copper (mg)	180	180
Cobalt (mg)	2	8.9
Dry matter (kg)	19	18.3

This ration costs £1.782/day and Table 9.2 compares its composition with the specified requirements. The nutrient requirements are binding restraints only for metabolisable energy, undegradable protein and copper. Several nutrients are in surplus supply – particularly magnesium, sodium and cobalt, which may be an undesirable nutritional imbalance in the diet. The situation could be improved by setting upper, as well as lower, limits for the level of supply of each of the nutrients. This approach, however, produces over-constrained problems with empty feasible sets in many cases. For instance, if, in this example, the supplies of sodium copper and cobalt are bound to upper limits of 50 g, 400 mg and 6 mg respectively, there is no solution that satisfies all the restrictions.

The reduction of surplus supplies of various nutrients in least-cost diets ought to be one of the goals in formulating rations using mathematical programming techniques. Besides cost and nutritional imbalances, another important consideration in ration formulation is how bulky the diet is. Obviously rations which are bulky, either because of low physical density or because of high moisture content, imply extra costs of storage and handling, and the limitations imposed by storage capacity; therefore, reducing the bulk of the ration becomes a desirable goal. With this in mind, Crabtree's example can now be reformulated as a problem with three different objectives: (1) minimise the cost of the ration; (2) minimise the imbalanced supply of copper, sodium and cobalt, and (3) minimise the bulk (i.e. the net weight) of the diet.

Obviously there is a situation of conflict in trying to achieve these objectives. The minimisation of cost is not compatible with minimising either the nutrient imbalances or the physical bulk of the ration. As conflicting objectives are being considered the problem cannot be solved through minimising each objective function subject just to the nutritional requirements.

Crabtree's example with multiple goals will be solved in what follows via GP and MOP techniques to illustrate how these techniques can be used fruitfully in ration formulation problems.

Ration formulation as a WGP problem

Let us now analyse the ration formulation problem with three objectives discussed in the preceding section and build it as a weighted goal programming (WGP) model. To do that we convert the three objectives into goals: that is, by fixing a target value for each objective and introducing corresponding negative and positive deviational variables and by using the dry matter content of the various ingredients from Table 9.3.

Table 9.3 Dry matter content of ingredients (g/kg)

Ingredient	Silage	Straw	Distillers' wet grains	Swedes	Barley	Dairy compound
Dry matter	270	820	258	108	833	800

Source: Crabtree (1982, p. 299)

Thus, the three goals are:

Goal g_1

The ration should not have a bulk larger than 40 kg of fresh weight and the expression for goal g_1 is given by:

$$3.704x_1 + 1.219x_2 + 3.876x_3 + 9.259x_4 + 1.200x_5 + 1.250x_6 + n_1 - p_1 = 40 \quad (9.1)$$

where the deviational variable n_1 measures the under-achievement of goal g_1 while p_1 plays the opposite role. As the desired level of ration bulk should not be greater than 40 kg, the deviational variable p_1 must be minimised.

Goal g_2

The supply of sodium in the diet should not exceed 100% of its specified requirements. This imbalance goal, I_{Na}, is obtained from constraint (7) in Table 9.1.

$$I_{Na} = 3.8x_1 + 1.1x_2 + 0.9x_3 + 2.6x_4 + 0.2x_5 + 5.4x_6 - 26.9 \quad (9.2)$$

Treating equation (9.2) as percentages instead of absolute values, we have:

$$\frac{(3.8x_1 + 1.1x_2 + 0.9x_3 + 2.6x_4 + 0.2x_5 + 5.4x_6 - 26.9)}{26.9} \cdot 100 \quad (9.3)$$

Therefore the expression for g_2 is:

$$14.13x_1 + 4.09x_2 + 3.35x_3 + 9.67x_4 + 0.74x_5 + 20.07x_6 + n_2 - p_2 = 200 \quad (9.4)$$

To achieve the desired level of this goal p_2 must be minimised.

Goals g_3 and g_4

Using the same procedure as for sodium, the goals for copper and cobalt are derived from constraints (8) and (9) in Table 9.1, so that their supplies, too, should not exceed 100% of the specified requirements. This is stated by the equations (9.5) and (9.6) below.

$$3.17x_1 + 1.78x_2 + 5.55x_3 + 2.11x_4 + 2.67x_5 + 10.55x_6 + n_3 - p_3 = 200 \quad (9.5)$$

$$4.5x_1 + 2x_2 + 10x_3 + 2x_4 + 2x_5 + 75x_6 + n_4 - p_4 = 200 \quad (9.6)$$

For achieving g_3 and g_4, p_3 and p_4 must be minimised.

Goal g_5

Lastly, the objective of minimising the cost can be converted into a goal by setting a target of £1.782 as cost, which corresponds to the minimum cost associated with the ration recommended by the ordinary LP approach. Therefore the expression for g_5 using the unit costs of ingredients from Table 9.1, is:

$$0.0482x_1 + 0.0268x_2 + 0.0698x_3 + 0.0926x_4$$
$$+ 0.1056x_5 + 0.1625x_6 + n_5 - p_5 = 1.782 \qquad (9.7)$$

To achieve g_5, p_5 is minimised.

The variables in the objective function have to be given in percentage deviations from the targets as minimisation of absolute deviations does not make sense when each goal is measured in different units. Hence, the elements of the objective function are standardised for the WPG model to give:

$$w_1 \frac{p_1}{40} \frac{100}{1} + w_2 \frac{p_2}{200} \frac{100}{1} + w_3 \frac{p_3}{200} \frac{100}{1} + w_4 \frac{p_4}{200} \frac{100}{1} + w_5 \frac{p_5}{1.782} \frac{100}{1} \qquad (9.8)$$

or

$$2.5w_1 p_1 + 0.50w_2 p_2 + 0.50w_3 p_3 + 0.50w_4 p_4 + 56.12w_5 p_5 \qquad (9.9)$$

Where $w_1, \ldots, w_5$ are the weights attached to the deviational variables representing the respective importance given to the achievement of the various goals. The structure of this WGP model is given in Table 9.4. Different solutions can be obtained by attaching different values to the w parameters. For instance, if $w_1 = w_2 = \ldots = w_5 = 1$, the simplex method provides the following optimum solution:

x_1 (silage) = 3.511 kg x_2 (straw) = 0.675 kg
x_3 (distillers' wet grain) = 3.80 kg x_4 (swedes) = 0
x_5 (barley) = 6.298 kg x_6 (dairy compound) = 4.716 kg

While the optimum values of the deviational variables are:

$n_1 = 0$ $p_1 = 1.477$ kg
$n_2 = 35.59\%$ $p_2 = 0$
$n_3 = 100\%$ $p_3 = 0$
$n_4 = 0$ $p_4 = 221.41\%$
$n_5 = 0$ $p_5 = 0.102$

Table 9.4 Weighted goal programming model for ration formulation

Objective function

Minimise: $2.5w_1p_1 + 0.50w_2p_2 + 0.50\ w_3p_3 + 0.50\ w_4p_4 + 56.12w_5p_5$

subject to

$3.704x_1 + 1.219x_2 + 3.876x_3 + 9.2594x_4 + 1.200x_5 + 1.250x_6 + n_1 - p_1 = 40$ (bulk)

$14.13x_1 + 4.09x_2 + 3.35x_3 + 9.67x_4 + 0.74x_5 + 20.07x_6 + n_2 - p_2 = 200$ (imbalance in sodium)

$3.17x_1 + 1.78x_2 + 5.55x_3 + 2.11x_4 + 2.67x_5 + 10.55x_6 + n_3 - p_3 = 200$ (imbalance in copper)

$4.5x_1 + 2x_2 + 10x_3 + 2x_4 + 2x_5 + 75x_6 + n_4 - p_4 = 200$ (imbalance in cobalt)

$0.0482x_1 + 0.0268x_2 + 0.0698x_3 + 0.09.26x_4 + 0.1056x_5 + 0.1625x_6 + n_5 - p_5 = 1.782$ (cost), and

$\underline{x} \in \underline{F}$ [technical constraints from Table 9.1 excluding sodium, copper and cobalt requirement restraints]

Table 9.5 Sets of weights used in the sensitivity analysis of the WGP solution

Run	Weight w_1 for g_1 (bulk)	Weight w_2 for g_2 (imbalance in sodium)	Weight w_3 for g_3 (imbalance in sodium)	Weight w_4 for g_4 (imbalance in sodium)	Weight w_5 for g_5 (cost)
1	1	1	1	1	1
2	1	2	2	2	1
3	2	1	1	1	1
4	1	1	1	1	2
5	1	3	3	3	1
6	3	1	1	1	1
7	1	1	1	1	3
8	1	1	1	1	4
9	1	2	2	2	4
10	2	1	1	1	4
11	1	1	1	1	5

This solution permits complete achievement of the goals g_2 (sodium supply imbalance) and g_3 (copper supply imbalance) by virtue of the fact that $p_1 = p_2 = 0$. The value of 1.477 kg for p_1 implies that goal g_1 has exceeded its target by 1.477 kg or the bulk of the ration is 41.477 kg. Similarly, $p_4 = 221.41\%$ means that the imbalance of cobalt has surpassed its target by that proportion; that is, the supply of this nutrient is 321.41% of the specified requirements. Finally, $p_5 = 0.102$ means that the cost of the ration is £0.102 over the set target of £1.782/day.

The sensitivity analysis of a WGP solution can provide very useful information for ration formulation. To demonstrate the generation and use of this information we have used eleven sets of weights, as shown in Table 9.5 and corresponding solutions are presented in Table 9.6. From this information it is possible to obtain something like a measure of the trade-offs between goals. Thus, if, for instance, the decision maker is undecided between solutions 1 and 2, the former will be preferred if the over-achievement of bulk by 2.486 kg can be traded off with a reduction in cost of £0.012 and for a drop in cobalt imbalance of 10.15%. Such an array of information giving trade-offs among various goals should enable the decision maker to choose the ration that best suits his aims. It must be pointed out that if the weight attached to the cost of the ration is given a value higher than 10, then the LP solution given above is obtained again.

Ration formulation as an LGP problem

To illustrate the use of lexicographic goal programming (LGP) in ration formulation, we have to establish the priority structure of goals in our example. Let us assume, for example, the goals of obtaining a ration costing less than £2 and weighing below 50 kg are situated in the first priority Q_1. These targets can be considered as minimum levels of achievement; that is, for the decision maker it is completely and absolutely necessary to use rations cheaper than £2 in order to obtain profits and with a bulk below 50 kg in order not to exceed the storage capacity.

Using equations (9.7) and (9.1) and on modifying the subscripts of the deviational variables, the goals in priority Q_1 are:

Cost:

$$0.0482x_1 + 0.0268x_2 + 0.0698x_3 + 0.0926x_4 + 0.1056x_5 + 0.1625x_6 + n_1 - p_1 = 2$$
$$(9.10)$$

Bulk:

$$3.704x_1 + 1.219x_2 + 3.876x_3 + 9.259x_4 + 1.200x_2 + 1.250x_6 + n_2 - p_2 = 50 \quad (9.11)$$

As regards weights attached to the achievement of the various goals, assume that formulating a ration costing less than £2 is twice as important as the attainment of required bulk. Now using the same procedure as for deriving the objective function of WGP the first component to be minimised in the lexicographic process is:

$$\frac{2 \times 100}{2} p_1 + \frac{100}{50} p_2 = 100p_1 + 2p_2$$

The next priority Q_2, is made up of goals g_2, g_3 and g_4 described above: that is, the supply-requirement imbalance for sodium, copper and cobalt should not be greater than 100%. This can be derived from equations (9.4), (9.7) and (9.8), with the appropriate subscripts for the deviational variables: thus the goals in priority Q_2 are:

Table 9.6 Sensitivity analysis of the WGP solution

Run	x_1 (silage) (kg)	x_2 (straw) (kg)	x_3 (distillers' wet grain (kg)	x_4 (swedes) (kg)	x_5 (barley) (kg)	x_6 (dairy compound)(kg)	Fresh weight (kg/day)		Surplus of Na as percent of requirement		Surplus of Cu as percent of requirement		Surplus of Co as percent of requirement		Cost (£)	
							Actual achievement of the goal	Deviation from the target	Actual achievement of the goal	Deviation from the target	Actual achievement of the goal	Deviation from the target	Actual achievement of the goal	Deviation from the target	Actual achievement of the goal	Deviation from the target
1	3.511	0.675	3.8	0	6.298	4.716	41.477	1.477	64.41	0	0	0	321.41	221.41	1.884	0.102
2	4.506	0	3.8	0	6.118	4.576	43.963	3.963	72.771	0	0	0	313.741	213.741	1.872	0.09
3	2.923	0.917	3.8	0	6.519	4.842	40	0	60.249	0	0.472	0	329.155	229.155	1.906	0.124
4	3.511	0.675	3.8	0	6.298	4.716	41.477	1.477	64.41	0	0	0	321.41	221.41	1.884	0.102
5	5.144	0	3.8	0	5.52	4.536	45.564	5.564	80.54	0	0	0	312.384	212.384	1.883	0.051
6	2.923	0.917	3.8	0	6.519	4.842	40	0	60.249	0	0	0	313.741	213.741	1.872	0.124
7	4.911	0	3.652	0	4.843	4.855	43.678	3.678	82.645	0	0	0	332.407	232.407	1.792	0.001
8	4.911	0	3.652	0	4.843	4.855	43.678	3.678	82.645	0	0	0	332.407	232.407	1.792	0.001
9	5.144	0	3.8	0	5.52	4.536	45.564	5.564	80.54	0	0	0	312.384	212.384	1.833	0.051
10	4.275	0.263	3.652	0	5.147	4.924	42.087	2.087	76.359	0	0	0	335.909	235.909	1.812	0.023
11	4.911	0	3.652	0	4.843	4.855	43.678	3.678	82.645	0	0	0	332.407	232.407	1.792	0.001

Sodium:

$$14.13x_1 + 4.09x_2 + 3.35x_3 + 9.67x_4 + 0.74x_5 + 20.07x_6 + n_3 - p_3 = 200 \qquad (9.12)$$

Copper:

$$3.17x_1 + 1.78x_2 + 5.55x_3 + 2.11x_4 + 2.67x_5 + 10.55x_6 + n_4 - p_4 = 200 \qquad (9.13)$$

Cobalt:

$$4.5x_1 + 2x_2 + 10x_3 + 2x_4 + 2x_5 + 75x_6 + n_5 - p_5 = 200 \qquad (9.14)$$

If the three goals in Q_2 are of equal importance, and as their targets and measurements are homogenous, then the second components to be minimised in the lexicographic process is given by $p_3 + p_4 + p_5$.

Finally, the last priority, Q_3, is made up of goals of finding a ration costing less than £1.782 with minimum bulk (which is 22.34 kg, see next section). It is interesting to note the similarity that appears to exist between goals in priorities Q_1 and Q_3. In fact, in Q_1 goals represent something like minimum levels of achievement while Q_3 represents desired levels of achievement. When goals are specified at both a minimum and desired level in a pre-emptive way, some authors call this approach two-stage LGP (Keown and Martin 1977).

The algebraic expression for the goals in Q_3 is derived from equations (9.1) and (9.7) as given below:

Cost:

$$0.0482x_1 + 0.0268x_2 + 0.0698x_3 + 0.0926x_4 + 0.1056x_5 + 0.1625x_6 + n_6 - p_6 = 1.782 \qquad (9.15)$$

Bulk:

$$3.704x_1 + 1.219x_2 + 3.876x_3 + 9.259x_4 + 1.200x_5 + 1.250x_6 + n_7 - p_7 = 22.34 \qquad (9.16)$$

Once again, assuming that the cost goal is twice as important as the ration bulk, the last component of the lexicographic minimisation process is given by:

$$\frac{2 \times 100}{1.782} p_6 + \frac{100}{22.34} p_7 = 112.2p_6 + 4.47p_7$$

The whole lexicographic minimisation process would be given by the following achievement function:

$$\text{Minimise } \underline{a} = [(100p_1 + 2p_2), (p_3 + p_4 + p_5), (112.2p_1 + 4.47p_2)] \tag{9.17}$$

The structure of our LGP model is given in Table 9.7.

Table 9.7 Lexicographic goal programming model for ration formulation

Achievement function: Minimise $\underline{a} = [(100p_1 + 2p_2), (p_3 + p_4 + p_5), (112.2p_1 + 4.47p_2)]$

subject to:

Q_1: $0.0482x_1 + 0.0268x_2 + 0.0698x_3 + 0.0926x_4 + 0.1056x_5 + 0.1625x_6 + n_1 - p_1 = 2$ (cost)
$3.704x_1 + 1.219x_2 + 3.876x_3 + 9.2594x_4 + 1.200x_5 + 1.250x_6 + n_2 - p_2 = 50$ (bulk)

Q_2: $14.13x_1 + 4.09x_2 + 3.35x_3 + 9.67x_4 + 0.74x_5 + 20.07x_6 + n_3 - p_3 = 200$ (imbalance in sodium)
$3.17x_1 + 1.78x_2 + 5.55x_3 + 2.11x_4 + 2.67x_5 + 10.55x_6 + n_4 - p_4 = 200$ (imbalance in copper)
$4.5x_1 + 2x_2 + 10x_3 + 2x_4 + 2x_5 + 75x_6 + n_5 - p_5 = 200$ (imbalance in cobalt)

Q_3: $0.0482x_1 + 0.0268x_2 ++ 0.0698x_3 + 0.0926x_4 + 0.1056x_5 + 0.1625x_6 + n_6 - p_6 = 1.782$ (cost)

i.e. cost

$3.704x_1 + 1.219x_2 + 3.876x_3 + 9.2594x_4 + 1.200x_5 + 1.250x_6 + n_7 - p_7 = 22.34$ i.e. bulk and

$\underline{x} \in \underline{F}$ [technical constraints from Table 9.1 excluding sodium, copper and cobalt requirement restraints]

On using the sequential linear method (see Chapter 3), we obtain the following optimum solution:

x_1 (silage) $= 5.145$ kg x_2 (straw) $= 0$
x_3 (distillers' wet grain) $= 3.80$ kg x_4 (swedes) $= 0$
x_5 (barley) $= 5.519$ kg x_6 (dairy compound) $= 4.536$ kg

While the optimum values of the deviational variables are:

$n_1 = £0.167$ $p_1 = 0$
$n_2 = 4.434$ kg $p_2 = 0$
$n_3 = 19.458\%$ $p_3 = 0$
$n_4 = 100\%$ $p_4 = 0$
$n_5 = 0$ $p_5 = 212.384\%$
$n_6 = 0$ $p_6 = 0.051$
$n_7 = 0$ $p_7 = 23.227$ kg

This solution permits complete achievement of the goals in Q_1; that is, the minimum levels set for cost and bulk are achieved. For Q_2 the targets for imbalances in sodium and copper supplies are also achieved but not for cobalt, which has a positive deviation of 212.384% which means that more than three times the requirement is supplied. Finally, in Q_3 the least cost and minimum bulk goals have positive deviations of £0.051 and 23.277 kg respectively; that is, their actual values are £1.833 and 45.517 kg.

Ration formulation as a MOP problem

To demonstrate the use of multiobjective programming (MOP) in ration formulation, assume that in Crabtree's problem the three objectives to be considered are: (1) minimise the cost; (2) minimise the bulk and (3) minimise the aggregated over-supply in sodium, copper and cobalt. The formal structure of the multiobjective model is thus given by:

$$\underline{Eff}\, Z(\underline{x}) = [z_1(\underline{x}), z_2(\underline{x}), z_3(\underline{x})] \tag{9.18}$$

where

$$z_1(\underline{x}) = 0.0482x_1 + 0.0268x_2 + 0.0698x_3 + 0.0926x_4 + 0.1056x_5 + 0.1625x_6 \text{ (cost (£))}$$

$$z_2(\underline{x}) = 3.704x_1 + 1.219x_2 + 3.876x_3 + 9.259x_4 + 1.200x_5 + 1.250x_6 \text{ (bulk (kg))}$$

$$z_3(\underline{x}) = 21.80x_1 + 7.87x_2 + 18.09x_3 + 13.78x_4 + 5.41x_5 + 105.62x_6 - 300$$
(imbalances in sodium, copper and cobalt)

subject to

$\underline{x} \in \underline{F}$ [technical constraints from Table 9.1 excluding sodium, copper and cobalt requirement restraints]

The expression for $Z_3(\underline{x})$ has been derived from the addition of the imbalances of the three nutrients as defined above.

For generating the efficient set we resort to the weighting method presented in Chapter 4. Applying this method six extreme efficient rations were generated. These rations are shown in Table 9.8.

It is interesting to note that solution 1 corresponds to the minimum cost ration, solution 3 to minimum imbalance and solution 4 to the minimum bulk ration. In other words, these three solutions establish the ideal or utopian point, which for our problem is: [£1.782; 382.399%; 22.334 kg].

Table 9.8 Multiobjective programming for ration formulations (efficient points)

Efficient solution	Silage x_1 (kg)	Straw x_2 (kg)	Distillers' wet grain x_3 (kg)	Swedes x_4 (kg)	Barley x_5 (kg)	Dairy Compound x_6 (kg)	Cost (£)	Bulk (kg)	Over-supply
1	4.981	0	3.343	2	2.778	5.069	1.782	59.385	451.499
2	5.007	0	3.555	0	4.815	4.884	1.792	43.655	418.265
3	3.569	0.386	3.8	0	6.566	4.679	1.901	41.618	382.399
4	0.395	3.260	0	0	4.502	10.118	2.226	22.344	827.306
5	3.306	0.494	3.8	0	6.665	4.736	1.911	40.959	384.006
6	2.855	0.945	3.8	0	6.544	4.856	1.908	39.830	389.823

The 'optimum' ration is one that is chosen by the decision maker from the set of efficient solutions. But, of course, the choice depends on the preferences of the decision maker, depending upon his subjective values attached to the trade-offs between cost, nutrient supply-requirement imbalance and bulk. For instance, if solution 1 (least cost ration) is preferred to solution 2, in that situation a reduction of 15.370 kg in bulk and 33.234% in imbalance does not compensate for an increase of £0.010 in cost. Alternatively, if the DM is not clear about the actual trade-offs then he could follow either the CP methods of Chapter 5 or the interactive approach as discussed in Chapter 6.

This chapter has demonstrated that the problem of diet formulation can be represented by GP and MOP far more realistically than is the case when ordinary LP is used. However, the problem that the specified nutrient requirements for a given animal ought to be met in a rigid and inflexible way still remains. However, we believe that a variation of GP incorporating penalty functions provides a better framework of analysis, as shown in the next chapter.

Note

1   This chapter is a slightly modified version of a paper by Rehman and Romero 1984.

Chapter ten
Livestock ration formulation via goal programming with penalty functions

It has already been pointed out that the process of formulating least cost diets using linear programming (LP) suffers unnecessarily from an over-rigid specification of nutritional and other requirements. Some relaxation of these rigidly imposed constraints would not seriously affect an animal's physical and economic performance. Small increases over and above the minimum cost of exactly meeting the specifications embodied in such restraints may achieve a mix of ingredients capable of yielding a superior performance that could in the long term more than recover the extra cost. The mathematically optimal solution from an LP model solves primarily a technical problem, which appears to have only a loose relationship with the economic problem of maximising the difference between costs and returns of feeding over time. The rigidity of the restraints built into an LP model is manifest in the fixed values assigned to the right-hand side parameters and the equal economic importance attached to each constraint.

In this chapter[1] we show how a GP model incorporating penalty functions can overcome, to a certain extent, the above problem, thus making the specification of minimum levels of nutrient requirements more flexible and realistic.

Penalty functions in diet formulation

Table 10.1 presents a modified form of a conventional LP model for formulating least-cost diet for Hereford X Friesian steers which has been adapted from France and Thornley (1984, pp. 49–52) to illustrate the methods being proposed in this chapter.

The example chosen is simple and could be criticised for considering only a limited number of feeds and for not presenting a realistic nutritional problem. As the main purpose of our work is to illustrate a methodological approach, the simplicity of the example cited can be forgiven.

Table 10.1 is self-explanatory. It should however, be noted that a minimum content of 10% maize ensures the palatability of the diet and avoids excessive acidity in the rumen. Further, the content of mineral-vitamin supplement is fixed to 1% for a sufficient supply of vitamins.

Table 10.1 France and Thornley's modified ration formulation problem

Barley x_1 (kg)	Maize x_2 (kg)	White Fish Meal x_3 (kg)	Soya bean x_4 (kg)	Mineral/ Vitamin Supplement x_5 (kg)	Salt x_6 (kg)			
10.52	14.275	24.5	14.4	59.56	11.5	=	Z min (pence/kg)	
1	1	1	1	1	1	=	1 weight (kg)	[1]
–	–	–	–	1	–	=	0.01 Mineral vitamin supplement (kg)	[2]
13.7	14.2	11.1	12.3	–	–	≥	13 Metabolisable energy – ME (MJ/kg)	[3]
108	98	701	503	–	–	≥	160 Crude protein – CP (g/kg)	[4]
0.5	0.2	79.3	2.3	120	–	≥	7 Calcium – Ca (g/kg)	[5]
3.8	2.7	43.7	10.2	60	–	≥	7 Phosphorus – P (g/kg)	[6]
0.2	0.1	16.1	5	60	400	≥	3 Sodium – Na (g/kg)	[7]
1.3	1	2.2	3.1	30	–	≥	2 Magnesium – Mg (g/kg)	[8]
–	1	–	–	–	–	≥	0.1 Maize (kg)	[9]

Source: France and Thornley (1984, pp. 49–52). Modifications consist of including salt as an ingredient and fixing the mineral vitamin supplement requirement\ to 1 per cent.

The optimal solution to this least-cost problem is:

x_1 (barley)   = 0.6117 kg x_2 (maize)   = 0.100 kg
x_3 (white fish meal)   = 0.0628 kg x_4 (soybean) = 0.2150 kg
x_5 (mineral-vitamin supplement) = 0.0100 kg x_6 (salt)   = 0.0005 kg

This diet costs 13.10 pence/kg. As the surplus variables for calcium, sodium, magnesium and maize content are zero, these requirements are the only binding restraints. The surplus supplies for energy, protein and phosphorus are 0.14 MJ/kg, 68.04 g/kg and 1.13 g/kg respectively. Thus this least-cost ration implies a moderate over-supply of phosphorus (about 16% over the minimum); however, the protein is included considerably in excess of the requirements (43% over the minimum), which could be undesirable nutritionally.

Instead of considering the nutrient requirements as rigid constraints we propose to treat them as goals, which may or may not be achieved. Hence the right-hand sides of the equations become targets that the decision maker tries to achieve as closely as possible but with the possibility of deviating from them. These goals are introduced to the original model by introducing a negative (n_i) and a positive (p_i) deviational variables, measuring the under- and over-achievements with respect to a specific nutritional requirement, for each constraint. However, as various requirements are specified in different units, some means of making the achievement of all the goals directly comparable is required, achieved simply by multiplying each constraint by 100 and then dividing the equation through by the right-hand side coefficient, in order to work in percentage terms. For instance, the energy constraint becomes the following goal:

$$105.38x_1 + 109.23x_2 + 85.38x_3 + 94.62x_4 + n_1 - p_1 = 100 \qquad (10.1)$$

The objective now is to minimise the sum of the deviations, $\Sigma(n_i + p_i)$, from 100% of the ith target. However, this formulation of the problem implies that relaxing a target by 1 per cent in either direction is as important as any other similar relaxation. Furthermore, any marginal change is of equal importance no matter how distant it is from the target; that is, it is equally significant when p_i changes either from 10 to 11 or from 1 to 2.

Both these assumptions are unrealistic. Thus to deal with the relative importance of nutrients included in a diet we resort to a weighted goal programming (WGP) approach, where the deviations are weighted according to the relative importance of each goal. For instance, if the negative deviation for energy (n_1) is twice more important than the negative deviation for protein (n_2), then the objective function will consist of the term $2n_1 + n_2$. Another possibility is to formulate the diet problem as a lexicographic goal programming (LGP) model, where the deviational variables of the different nutrients are minimised in a pre-emptive way. We shall discuss this model in a later section of this chapter.

The problem of specifying relative relaxation of a given nutritional restraint in diet formulation can be tackled by using a penalty system so that an animal's requirements are kept within nutritionally desirable limits. When these limits are violated, a penalty scale operates, therefore, the objective function measures the total penalty incurred due to the violations of various nutritional requirements that were set in the model.

To illustrate the operation of this system we have assumed the penalty scales shown in Table 10.2 for the various nutritional requirements that were specified in France and Thornley's example.

These penalty scales imply the inclusion of the functions of the type shown in Figures 10.1 to 10.3. Such functions can be built in a GP model using some of the methods proposed in the literature by Kvanli (1980), Can and Houck (1984) and Romero (1984). The method proposed by Can and Houck has some operational advantages over others, particularly when marginal penalty functions are monotonically increasing with respect to the targets, as in our problem. Thus the following equations illustrate the derivation of restraints representing the penalty functions for calcium when Can and Houck's method is used.

$$7.14x_1 + 2.86x_2 + 1132.86x_3 + 32.85x_4 + 1714.29x_5 + n_{31} + n_{33} - p_{31} - p_{32} = 120 \quad (10.2)$$
$$0 \leq n_{31} \leq 20, 0 \leq n_{32} \leq 10, 0 \leq n_{33} \leq 20, 0 \leq p_{31} \leq 40, 0 \leq p_{32} \leq 40$$

Table 10.2 Penalty scales for the nutrient requirements in France and Thornley's model

	Units	Marginal penalty
Metabolisable energy	Below 90%	Infinite
Crude protein	90–100%	4
	100–110%	0
	110–120%	2
	Over 120%	Infinite
Sodium Magnesium	Below 70%	Infinite
Calcium Phosphorus	70–90%	4
	90–100%	1
	100–120%	0
	120–160%	1
	160–200%	4
	Over 200%	Infinite
Maize	Below 60%	Infinite
	60–90%	4
	90–100%	1
	Over 100%	0

The negative deviational variables n_{31}, n_{32} and n_{33} measure the percentage points of calcium content ranging between 70–90%, 90–100% and 100–120% respectively. Similarly, the deviational variables p_{31} and p_{32} measure the percentage points of calcium contents ranging from 120% to 160% and 160% to 200% respectively. The contributions of these variables to the objective function of the GP model are given by $4n_{31} + n_{32} + p_{31} + 4p_{32}$. The penalty goals and constraints for other nutrient requirements are built similarly using the scales given in Table 10.2.

Diet formulation as a WGP model with penalty functions

Table 10.3 gives the structure of the weighted goal programming (WGP) model incorporating the penalty functions derived from the scales given in Table 10.2. All the constraints of the original problem have been relaxed, except for the weight of the mix and the mineral-vitamin supplement content; therefore these two must be treated as rigid constraints.

The objective function of the model measures the total penalty generated by the feed mix. The coefficients w_i ($i = 1, ..., 7$) measure the relative importance attached to the

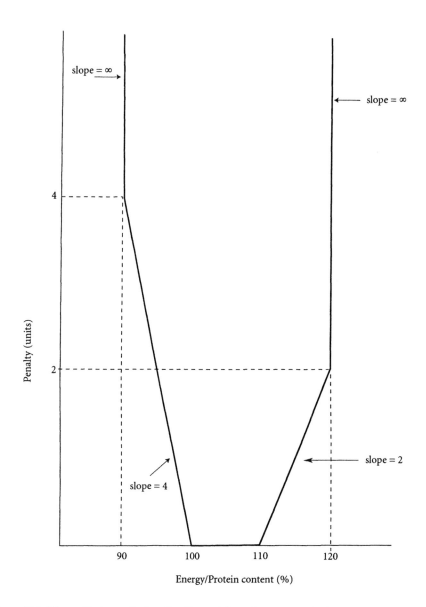

Figure 10.1 Three sided penalty function for protein and energy content

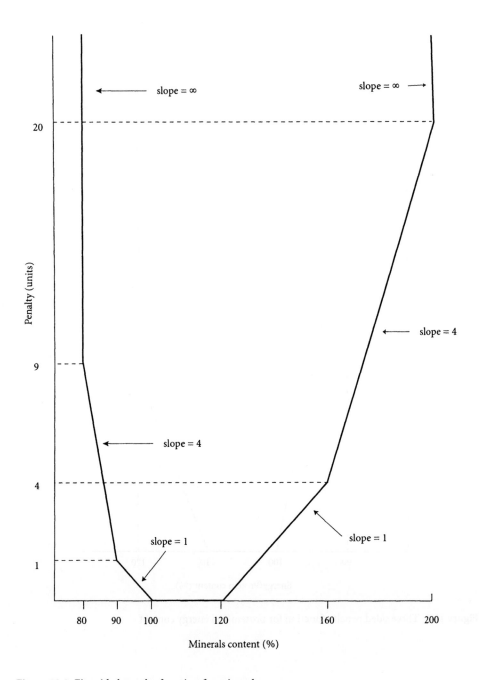

Figure 10.2 Five sided penalty function for minerals

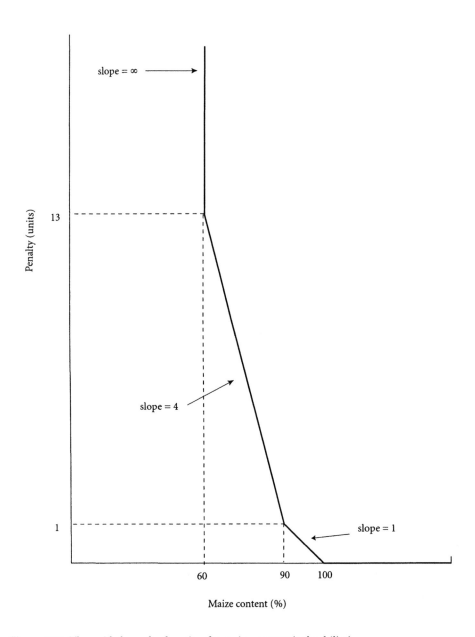

Figure 10.3 Three sided penalty function for maize content (palatability)

Table 10.3 Weighted goal programming mode with penalty functions for diet formulation

Objective function

$$4w_1n_{11} + 2w_1p_{11} + 4w_2n_{21} + 2w_2p_{21} + 4w_3n_{31} + w_3n_{32} + w_3p_{31} + 4w_3p_{32} + 4w_4n_{41}$$
$$+ w_4n_{42} + w_4p_{41} + 4w_4p_{42} + 4w_5n_{51} + w_5n_{52} + w_5p_{51} + 4w_5p_{52} + 4w_6n_{61} + w_6n_{62}$$
$$+ w_6p_{61} + 4w_6p_{62} + 4w_7n_{71} + w_7n_{72}$$

subject to

$$x_1 + x_2 + x_3 + x_4 + x_5 + x_6 = 1 \qquad (1) \quad \text{Weight (kg)}$$

$$x_5 = 0.01 \qquad (2) \quad \text{Supplement (kg)}$$

$$105.38x_1 + 109.23x_2 + 85.38x_3 + 94.62x_4 + n_{11} + n_{12} - p_{11} = 110 \qquad (3) \quad \text{Metabolisable energy}$$
$$0 \leq n_{11} \leq 10, \; 0 \leq n_{12} \leq 10, \; 0 \leq p_{11} \leq 10$$

$$67.5x_1 + 61.25x_2 + 438.13x3 + 314.38x_4 + n_{21} + n_{22} - p_{21} = 110 \qquad (4) \quad \text{Crude protein (\%)}$$
$$0 \leq n_{21} \leq 10, \; 0 \leq n_{22} \leq 10, \; 0 \leq p_{21} \leq 10$$

$$7.14x_1 + 2.86x_2 + 1132.86x3 + 32.85x_4 + 1714.29x_5 + n_{31} + n_{32} + n_{33} - p_{31} - p_{32} = 120 \qquad (5) \quad \text{Calcium (\%)}$$
$$0 \leq n_{31} \leq 20, \; 0 \leq n_{32} \leq 10, \; 0 \leq n_{33} \leq 20, \; 0 \leq p_{21} \leq 40, \; 0 \leq p_{32} \leq 40$$

$$54.29x_1 + 38.57x_2 + 624.29x_3 + 145.71x_4 + 857.14x_5 + n_{41} + n_{42} + n_{43} - p_{41} - p_{42} = 120 \qquad (6) \quad \text{Phosphorus (\%)}$$
$$0 \leq n_{41} \leq 20, \; 0 \leq n_{42} \leq 10, \; 0 \leq n_{43} \leq 20, \; 0 \leq p_{41} \leq 40, \; 0 \leq p_{42} \leq 40$$

$$6.66x_1 + 3.33x_2 + 536.66x_3 + 166.66x_4 + 2000x_5 + 13333x_6 + n_{51} + n_{52} + n_{53} - p_{51} - p_{52} = 120 \qquad (7) \quad \text{Sodium (\%)}$$
$$0 \leq n_{51} \leq 20, \; 0 \leq n_{52} \leq 10, \; 0 \leq n_{53} \leq 20, \; 0 \leq p_{51} \leq 40, \; 0 \leq p_{52} \leq 40$$

$$65x_1 + 50x_2 + 110x_3 + 155x_4 + 1500x_5 + n_{61} + n_{62} + n_{63} - p_{61} - p_{62} = 120 \qquad (8) \quad \text{Magnesium (\%)}$$
$$0 \leq n_{61} \leq 20, \; 0 \leq n_{62} \leq 10, \; 0 \leq n_{63} \leq 20, \; 0 \leq p_{62} \leq 40$$

$$1000x_1 + n_{71} + n_{72} - p_{71} = 100 \qquad (9) \quad \text{Maize (\%)}$$
$$0 \leq n_{71} \leq 30, \; 0 \leq p_{71} \leq 10$$

$$10.52x_1 + 14.27x_2 + 24.50x_3 + 14.10x_4 + 59.56x_5 + 11.50x_6 \leq T \qquad \text{(Parametric restraint: cost (pence/kg))}$$

$$x \geq 0, \, n \geq 0, \, p \geq 0$$

achievement of different targets. Thus w_1 indicates the importance of the energy target; while w_2 does the same for protein and so on. The cost of the mix is treated as a parametric restraint. Thus, by varying its value, several diets minimising the total penalty will be obtained.

Mathematically the model in Table 10.3 is a conventional parametric LP problem that can be solved by the simplex method and the solutions obtained for different sets of weights and values of the cost parameter are given in Table 10.4. The least-cost value of 13.10 pence/kg provides the starting point. The set of first six runs attached the same importance to the different nutrient-goals by setting $w_1 = w_2 = \ldots = w_7$. It was not possible to find a feasible mix whatever total penalty is incurred below a cost of 11.95 pence/kg.

A sensitivity analysis with weights w_i provides useful information. Thus in our example, when the weights for energy and protein goals are two and four times the weight given to every other goal, then there are significant changes in the optimum solution for some levels of the cost (see rows 7 and 8 of Table 10.4). Similarly, when the weight attached for the palatability goal is three times the weight for any other goal then there is a change in the optimum solution for a cost level of 12.20 pence/kg of the mix.

Despite the straightforward nature of the example used for illustration, these results are remarkable. On comparing solutions 1, 2 and 3 with the least-cost diet, it is easily seen that it is possible to reduce the cost of the mix considerably, without impairing its nutritional qualities. Solution 1 provides a ration that is 0.48 pence/kg cheaper than the least-cost diet, and also offers a more balanced diet. The surplus supply of protein in the least-cost solution (about 68 g/kg) is reduced considerably when the penalty system is used. Thus it is not unreasonable to assume that the feed mixer would prefer solution 1 to the least-cost ration; he may want to trade 0.24 g/kg deficit in magnesium content for 0.48 pence/kg to achieve a more balanced supply of protein in the diet. Similarly, solutions 2 and 3 offer important reductions in the cost of the mix and a more balanced diet for a small deficit in the calcium and phosphorus content.

In fact, Table 10.4 can be treated as a 'choice set' for the feed mixer which allows him to choose the 'best' solution according to the subjective trade-offs between different deficits of various nutrients and several cost levels. For example, solution 6 should be preferred by the feed mixer to the least-cost diet if he accepts a trade-off of 1.15 pence/kg of cost of the diet for the following deficits: 16 g/kg of protein, 2.10 g/kg of calcium, 0.89 g/kg of phosphorus, 0.33 g/kg of magnesium and 38.20 g/kg of maize (palatability).

Diet formulation as an LGP model with penalty functions

As stated earlier, one possible approach to explicit recognition of the relative importance of nutritional goals is to formulate the problem as an LGP model. To illustrate how the LGP model with penalty functions can be used to find optimal diets, we assume that the following priority structure of goals has been postulated for our example problem.

Table 10.4 Solutions for the weighted goal programming model incorporating penalty functions

Weights	Maximum Cost pence allowed	Barley x_1 (kg)	Maize x_2 (kg)	Fish x_3 (kg)	Soya x_4 (kg)	Min-Vit x_5 (kg)	Salt x_6 (kg)	Metabolizable energy %	MJ/kg	Crude Protein %	g/kg	Calcium %	g/kg	Phosphorus %	g/kg	Sodium %	g/kg	Magnesium %	g/kg	Maize palatability %	g/kg
												Actual achievements of goals specified									
$w_1 = w_2 = \dots = w_7$	13.1	0.7431	0.1	0.0658	0.0789	0.01	0.0027	102.34	13.30	110	176	100	7	105.43	7.38	103.08	3.09	87.8	1.76	100	100
$w_1 = w_2 = \dots = w_7$	12.5	0.7495	0.1	0.0569	0.0811	0.01	0.0023	102.46	13.32	107.22	172	90	6.3	100.53	7.04	100	3	87.58	1.75	100	100
$w_1 = w_2 = \dots = w_7$	12.4	0.7792	0.1	0.0583	0.0498	0.01	0.0027	102.74	13.36	100	160	90.77	6.35	98.46	6.89	101.11	3	84.82	1.70	100	100
$w_1 = w_2 = \dots = w_7$	12.2	0.8290	0.09	0.0574	0.0102	0.01	0.0031	103.10	13.40	90	144	89.06	6.23	94.59	6.62	100	3.03	81.34	1.63	90	90
$w_1 = w_2 = \dots = w_7$	12	0.8350	0.0751	0.0401	0.0364	0.01	0.0035	103.07	13.40	90	144	70	4.9	87.17	6.1	100	3	83.01	1.66	75.1	75.1
$w_1 = w_2 = \dots = w_7$	11.95	0.8488	0.0618	0.0401	0.0361	0.01	0.0035	103.04	13.44	90	144	70	4.9	87.34	6.11	100	3	83.27	1.67	61.8	61.8
$w_1 = w_2 = 2; w_3 = \dots = w_7 = 1$	12.2	0.7811	0.09	0.0389	0.0767	0.01	0.0031	102.76	13.36	99.58	159	70	4.9	90.16	6.31	100	3	86.49	1.73	90	90
$w_1 = w_2 = 4; w_3 = \dots = w_7 = 1$	12	0.8345	0.06	0.0397	0.0526	0.01	0.0033	102.86	13.37	93.95	150	70	4.9	88.65	6.2	100	3	84.77	1.69	60	60
$w_1 = \dots w_6 = 1; w_7 = 3$	12.2	0.8163	0.1	0.0526	0.0176	0.01	0.0033	103.14	13.41	90	114	83.85	5.87	92.4	6.47	101.01	3.03	81.63	1.63	100	100

First priority – Q_1: To obtain a mix cheaper than the least-cost solution; that is less than 13.10 pence/kg. Using the coefficients from the objective function in Table 10.1, the goal making up priority Q_1 is given by:

$$10.52x_1 + 14.27x_2 + 24.50x_3 + 14.10x_4 + 59.56x_5 + 11.50x_6 + n_1 - p_1 = 13.10$$

To achieve this goal the deviational variable p_1 must be minimised.

Second priority – Q_2: To minimise the total penalty incurred in terms of imbalances in energy and protein supplies and the palatability of the diet. The goals comprising priority Q_2 would be given by the expressions (3) for energy, (4) for protein and (9) for palatability from Table 10.3. To achieve these goals the sum $4n_{11} + 2p_{11} + 4n_{21} + 2p_{21} + 4n_{71} + n_{72}$ must be minimised.

Third priority – Q_3: To minimise the total penalty incurred in terms of minerals content and to minimise the cost of the mix. It is interesting to note that in priority Q_1 the cost goal represents something like a minimum level of achievement corresponding to the least-cost solution, while in Q_3 the same goal becomes a desired level of achievement. These goals are given by expressions (5), (6), (7) and (8) from Table 10.3. In order to establish the desired cost goal, an infeasibly low value of 11.50 pence/kg was set. To make the goals of Q_3 comparable the desired cost goal is expressed in percentage terms as below:

$$91.48x_1 + 124.09x_2 + 213.04x_3 + 125.22x_4 + 517.91x_5 + 100x_6 + n_2 - p_2 = 100$$

To achieve the priority Q_3 the sum of deviational variables to be minimised is:

$$4n_{31} + n_{32} + 4p_{32} + 4n_{41} + n_{42} + p_{41} + 4p_{42} + 4n_{51} + n_{52} + p_{51} + 4p_{52} + 4n_{61}$$
$$+ n_{62} + p_{61} + 4p_{62} + wp_2$$

where w represents the weight attached to the desired cost goal.

Table 10.5 shows the whole structure of the LGP model with penalty functions for France and Thornley's example. Once again using the sequential linear method and giving the weight w a value of 10, the following solution is obtained:

x_1 (barley)	= 0.7522 kg	x_2 (maize)	= 0.1000 kg
x_3 (white fish meal)	= 0.0571 kg	x_4 (soybean)	= 0.0778 kg
x_5 (mineral-vitamin supplement)	= 0.0100 kg	x_6 (salt)	= 0.0029 kg

This solution permits complete achievement of the goals making up priority Q_1 (the minimum level of 13.10 pence/kg for the cost) and Q_2 (the targets for energy, protein and palatability are achieved without incurring any penalty). However, goals making up priority

Q_3 are not fully achieved. Thus, there is a deviation in the cost of the mix of 0.99 pence/kg with respect to the low infeasible target of 11.50 pence/kg (the actual cost is 12.49 pence/kg). The minerals content targets are achieved without any penalty except for calcium (with an actual achievement of 6.3 g/kg) and for magnesium (with an actual achievement of 1.74 g/kg). It is interesting to point out the similarity between this solution and solution 2 of Table 10.4.

A sensitivity analysis by rearranging the order of priorities, or changing the value of some weights as the parameter w attached to the desired cost goal can produce useful information for the feed mixer.

Table 10.5 Lexicographic goal programming model with penalty functions for diet formulation

Achievement function:

$$\text{Minimise } \underline{a} = [(p_1), (4n_{11} + 2p_{11} + 4n_{21} + 2\,p_{21} + 4n_{71} + n_{72}), (4n_{31} + n_{32} + p_{31} + 4p_{32} + 4n_{41} + n_{42}$$
$$+ \, p_{41} + 4p_{42} + 4n_{51} + n_{52} + p_{51} + 4p_{52} + 4\,n_{61} + n_{62} + p_{61} + 4p_{62} + 10p_2)]$$

subject to

Q_1: { $10.52x_1 + 14.27x_2 + 24.50x_3 + 14.40x_4 + 59.56x_5 + 11.50x_6 + n_1 - p_1 = 13.10$
(cost: minimum level of achievement)

Q_2: { goals and constraints (3), (4) and (9) from Table 10.3 (energy, protein and palatability)

Q_3: { goals and constraints (5), (6), (7) and (8) from table 10.3 (minerals)

$91.48x_1 + 124.09x_2 + 213.04x_3 + 125.22x_4 + 517.91x_5 + 100x_6 + n_2 - p_2 = 100$ (cost: desired level of achievement)

$\underline{x} \geq \underline{0}, \underline{n} \geq \underline{0}, \underline{p} \geq \underline{0}$

An assessment

In this chapter we have attempted to explore the possibility of treating the livestock diet formulation problem as a GP model linked to a system of penalty scales. In our view this approach represents a new departure from the traditional LP paradigm. Notwithstanding the use of only one example, the results are encouraging. In fact with this method, once the rigidity of the nutritional specification has been overcome, it seems possible to reduce considerably the cost of the diet without jeopardising its nutritional quality.

We must however offer some comments regarding the potential application of GP with penalty functions to ration formulation even at this exploratory stage. With the proposed approach, the size of the model is considerably larger than when the problem is formulated as an ordinary LP model. Given the availability of increasingly sophisticated software packages for solving large LP problems, this should not pose a real problem. However, if the size of the matrix is a problem, then the situation could be mitigated in two possible ways. First, by using solution algorithms that do not treat bounds on variables as constraints and hence the computing time requirements of the GP model would be similar to a corresponding LP formulation. Second, the construction of penalty functions could be confined to a selected and small set of nutrients. For this purpose it would be a good starting point to calculate the shadow prices of the several requirements. In fact, constructing penalty functions for nutrient requirements that have high shadow prices will have a considerable impact on the cost of the diet. On the other hand, this will not be the case for nutrient requirement with low or nil shadow prices and hence there is no need to construct penalty functions for such restraints.

Finally, a critical aspect of the proposed approach is how the penalty scales are actually arrived at. As our purpose was mainly to establish a framework for finding optimal livestock diets and to demonstrate its functioning, the penalty scales used here were determined arbitrarily without absolute regard to their accuracy and nutritional feasibility. Clearly, prior to practical implementation of this approach, research into how livestock performance responds to incremental changes in the intake of various nutrients around the minimum specification is needed.

Note

1   This chapter is a slightly modified version of a paper by Rehman and Romero (1987b).

Chapter eleven

Optimum fertiliser use via goal programming with penalty functions

This chapter[1] is an extension of the previous one and the purpose here is to show the potential of goal programming (GP) with penalty functions for solving problems other than livestock ration formulation. In this chapter it is shown how this approach is applied to establishing an optimum fertiliser combination for sugar beet grown in the western part of Andalusia, Spain.

The crop is sown in autumn and harvested in mid-summer. Under irrigated conditions yields of 60 t/ha are obtained easily. To achieve this level of yield as a goal for the Alfisol soils in the Guadalquivir valley the following lower and upper limits for fertiliser requirements are recommended: 160–180 kg/ha of nitrogen (N), 60–80 kg/ha of phosphorus (P_2O_5), 80–100 kg/ha of potassium (K_2O), 1–1.3 kg/ha of boron (B), 0.5–1 kg/ha of manganese (Mn) and 0.15–0.25 kg/ha of zinc (Zn).[2] It is necessary to set upper limits on the supply of nutrients due to sugar beet's sensitivity to excessive levels.

The fertiliser compounds that can be used and their cost and composition are specified in Table 11.1. This table represents the LP matrix for finding the least-cost combination of fertilisers. The last three rows of the matrix have been included to ensure that a maximum of 35% of the requirements for the primary nutrients are supplied during the spring through the use of the last four compounds that are foliar fertilisers.

The solution to this least-cost fertiliser combination problem is:

x_1 (46-0-0)   = 288.19 kg x_2 (0-45-0)   = 10.12kg
x_4 (9-18-27B)   = 300.61 kg x_6 (FERTILUQ) =   1.67 kg
x_7 (UTEFOL-FICOOP) =   3.33 kg

The cost of this combination is 23,200 pta/ha. For this solution the lower constraints for nitrogen, phosphorus and boron are binding and the upper constraints for manganese and zinc are also binding, that is, the supply of nitrogen, phosphorus and boron provided by the least-cost solution coincides exactly with their minimum requirements, while the supply of manganese and zinc coincides exactly with the maximum requirements.

Table 11.1 Least-cost fertiliser mixture – basic matrix

Urea 46-0-0 x_1 (kg)	Superphosphate 0-45-0 x_2 (kg)	8-24-8 x_3 (kg)	9-18-27B x_4 (kg)	LAIFOL 0-9-18 x_5 (kg)	FERTILUQ 0-35-35 x_6 (kg)	UTERPOL FICOOP 6-18-27 x_7 (kg)	QIMIFOL E (ERT) 8-8-8 x_8 (kg)			
38.45	32.4	31.25	34.92	376	412	174	275	=	Z min (pta/kg)	
0.46		0.08	0.09	0.09	0.35	0.06	0.08	≥	160 nitrogen (N)	[1]
	0.45	0.24	0.18	0.18	0.35	0.18	0.08	≥	60 phosphate (P_2O_5)	[2]
		0.08	0.27		0.02	0.27	0.08	≥	80 potassium (K_2O)	[3]
			0.003		0.50	0.02	0.025	≥	1 boron (B)	[4]
				0.02	0.50	0.05	0.05	≥	0.5 manganese (Mn)	[5]
				0.01	0.05	0.05	0.10	≥	0.15 zinc (Zn)	[6]
0.46		0.08	0.09		0.35	0.06	0.08	≤	180 nitrogen (N)	[7]
	0.45	0.24	0.18	0.09	0.35	0.18	0.08	≤	80 phosphate (P_2O_5)	[8]
		0.08	0.27	0.18	0.02	0.27	0.08	≤	100 potassium (K_2O)	[9]
			0.003		0.50	0.02	0.025	≤	1.3 boron (B)	[10]
				0.02	0.50	0.05	0.05	≤	1 manganese (Mn)	[11]
				0.01	0.05	0.05	0.10	≤	0.25 zinc (Zn)	[12]
-16.1		-2.80	-3.15			3.90	5.20	≤	0	[13]
	-15.75	-8.40	-6.30	5.85	22.75	11.70	5.20	≤	0	[14]
		-2.80	-9.45	11.70	22.75	17.55	5.20	≤	0	[15]

Table 11.2 Penalty scales for the nutrient requirements for sugar cane

	Units	Marginal penalty
Nitrogen (N)	Below 80%	Infinite
	80–90%	4
	90–100%	1
	100–110%	0
	110–120%	3
	Over 120%	Infinite
Phosphorus (P_2O_5)	Below 70%	Infinite
	70 -100%	2
	100–130%	0
	130–150%	2
	Over 150%	Infinite
Potassium (K_2O)	Below 60%	Infinite
	60–80%	4
	80–100%	1
	100–110%	0
	110–150%	2
	Over 150%	Infinite
Boron (B)	Below 70%	Infinite
	70–80%	4
	80–100%	2
	100–120%	0
	120–130%	1
	130–140%	4
	Over 140%	Infinite
Manganese (Mn)	Below 60%	Infinite
	60–80%	4
	80–100%	2
	100–150%	0
	150–190%	1
	190–220%	2
	Over 220%	Infinite
Zinc (Zn)	Below 60%	Infinite
	60–80%	3
	80–100%	1
	100–150%	0
	150–200%	2
	Over 200%	Infinite

Table 11.3 Sensitivity analysis of the GP model with penalty functions

Set of Weights	Cost pstas	x_1 (kg)	x_2 (kg)	x_3 (kg)	x_4 (kg)	x_6 (kg)	x_7 (kg)	Nitrogen (N) Actual achievement %	kg	Phosphorus (P_2O_5) Actual achievement %	kg	Potassium (K_2O) Actual achievement %	kg	Boron (B) Actual achievement %	kg	Manganese (Mn) Actual achievement %	kg	Zinc (Zn) Actual achievement %	kg
$w_1 = w_2 = \ldots = w_6 = 1$	23,200	283.8	–	23.5	303.3	0.61	3.9	100	160	101.9	61.15	106.3	85.04	100	1	100	0.5	150	0.225
$w_1 = w_2 = \ldots = w_6 = 1$	22,000	256.6	–	18.7	303.3	0.61	3.9	91.9	147.1	100	60	105.8	84.66	100	1	100	0.5	150	0.225
$w_1 = w_2 = \ldots = w_6 = 1$	21,000	237.3	–	35.3	281.2	0.61	3.9	86	137.6	100	60	100	80	93.3	0.933	100	0.5	150	0.225
$w_1 = w_2 = \ldots = w_6 = 1$	20,000	218	–	52	258.9	0.61	3.9	80	128	100	60	94.15	75.32	86.7	0.867	100	0.5	150	0.225
$w_1 = w_2 = \ldots = w_6 = 1$	19,000	225.1	–	36.1	236.7	0.61	3.9	80	128	86.95	52.17	85.06	68.05	80	0.800	100	0.5	150	0.225
$w_1 = w_2 = \ldots = w_6 = 1$	18,000	232.2	–	–	232.5	0.61	3.9	80	128	71.26	42.76	80	64	78.7	0.787	100	0.5	150	0.225
$w_1 = w_2 = \ldots = w_6 = 1$	17,500	235.8	5.79	–	215.1	0.78	2.2	80	128	70	42	73.7	58.96	70.5	0.705	100	0.5	100	0.150
$w_1 = w_2 = w_3 = 2; w_4 = w_5 = w_6 = 1$	23,200	284.8	–	18.3	303.3	1.13	3.4	100	160	100	60	105.8	84.66	100	1	146.8	0.734	150	0.225
$w_1 = w_2 = w_3 = 2; w_4 = w_5 = w_6 = 1$	21,000	241.8	–	35.2	282.6	0.78	2.2	87.3	139.6	100	60	100	80	90.8	0.908	100	0.5	100	0.150
$w_1 = w_2 = w_3 = 2; w_4 = w_5 = w_6 = 1$	18,000	230.9	–	–	240.2	0.78	2.2	80	128	73.2	43.92	82.17	65.74	78.1	0.781	100	0.5	100	0.150
$w_1 = w_2 = w_3 = 3; w_4 = w_5 = w_6 = 1$	17,500	234.9	3.84	–	220.6	0.84	1.6	80	128	70	42	75.33	60.26	71	0.710	100	0.5	80	0.120
$w_1 = w_2 = w_3 = 4; w_4 = w_5 = w_6 = 1$	19,000	229.7	25.4	–	246.7	0.78	2.2	80	128	94.2	56.52	84.34	67.47	80	0.800	100	0.5	100	0.150
$w_1 = w_2 = w_3 = 1; w_4 = w_5 = w_6 = 2$	19,000	225.1	–	–	269	0.61	3.9	80	128	82.22	49.33	92.37	73.9	89.7	0.897	100	0.5	150	0.225
$w_1 = w_4 = 2; w_2 = \ldots = w_6 = 1$	20,000	230.5	–	–	291.6	0.61	3.9	82.8	132.5	89.01	53.41	100	80	96.5	0.965	100	0.5	150	0.225

In this case study it should be interesting to examine the effects of relaxing certain binding constraints on the cost of the fertiliser combination by using GP with penalty functions. For this purpose the penalty scales shown in Table 11.2 have been assumed for the different soil nutrient requirements in our problem. The constraints corresponding to these penalty functions are easily derived using the method outlined in Chapter 10. Thus, for the nitrogen requirements, once the goal has been standardised (i.e. expressed in percentage terms), the following equations illustrate the restraints associated with the penalty functions for this nutrient:

$$0.287x_1 + 0.05x_3 + 0.056x_4 + 0.037x_7 + 0.05x_8 + n_1 + n_2 + n_3 - p_4 = 110$$
$$0 \le n_1 \le 10, 0 \le n_2 \le 10, 0 \le n_3 \le 10, 0 \le p_4 \le 10$$

where

$4n_1 + n_2 + 3p_4$ is the objective function of the GP model.

The objective function of the model measures the total penalty generated by the fertiliser combination. In order to discriminate the relative importance given to the achievement of the different targets, a set of weights (w_1, w_2, ..., w_6) is attached to the deviational variables for the different nutrients that appear in the objective function. Thus w_1 measures the importance of the target nitrogen, w_2 of the target phosphorus, and so on, so the contribution to the objective function of the model for nitrogen will be given by w_1 ($4n_1 + n_2 + 3p_4$).

The cost of the combination is treated as a parametric restraint as the right-hand side of this inequality (cost of the combination) is a parameter; thus, by varying its value several fertiliser combinations that minimise the total penalty can be obtained.

Table 11.3 shows the solutions generated for different sets of weights and for different values of the fertiliser combination cost. The least-cost combination (23,200 pta/ha) provided the starting point. In the first seven computer runs the same importance was attached to the different nutrient-goals (i.e. setting $w_1 = w_2 = ... = w_6$). Below a cost of 17,500 pta/ha it was not possible to find any feasible fertiliser combination for any total penalty incurred. In order to furnish the farmer with useful complementary information, a sensitivity analysis with the weights was carried out to generate the last seven solutions.

Table 11.3 provides information that a farmer can use to choose the 'best solution' depending on his subjective trade-offs among deficits of different nutrients and costs. For instance, solution 3 is likely to be chosen in preference to the least-cost solution if the farmer is willing to trade 2,200 pta/ha of the cost of the fertiliser combination for a deficit of 22.45 kg/ha of nitrogen and 0.067 kg/ha of boron. This trade-off will be profitable if the market value of the fall in crop yield due to the lack of nitrogen and boron is less than 2,200 pta/ha.

Notes

1 This chapter summarises the practical aspects of the paper by Minguez et al. (1987).

2 Personal communications with L. Gordo from AIMCRA (Society for Research and Breeding of Sugar Beet) and P. Gonzalez from Instituto Nacional de Investigaciones Agrarias, 1984.

References

Agrawal, R.C. and Heady, E.O. (1968). Application of game theory models in agriculture. *Journal of Agricultural Economics*, **19**, 207–218.

Alphonce, C. B. (1997). Application of the Analytic Hierarchy Process in agriculture in developing countries. *Agricultural Systems*, **53**, 97–112.

Amador, F. and Romero, C. (1989). Redundancy in lexicographic goal programming: an empirical approach. *European Journal of Operational Research*, **41**, 347–354.

Apland, J., Barnes, R.N. and Justus, F. (1984). The farm lease: an analysis of owner-tenant and landlord preferences under risk. *American Journal of Agricultural Economics*, **66**, 376–384.

Arthur, J.L. and Ravindran, A. (1978). An efficient goal programming algorithm using constraint partitioning and variable elimination. *Management Science*, **24**, 867–868.

Arthur, J.L. and Ravindran, A. (1980a). PAGP, A partitioning algorithm for (linear) goal programming problems. *ACM Transactions Mathematical Software*, **6**, 378–386.

Arthur, J.L. and Ravindran, A. (1980b). A branch and bound algorithm with constraint partitioning for integer goal programming problems. *European Journal of Operational Research*, **4**, 421–425.

Balachandran, M. and Gero, J.S. (1984). A comparison of three methods for generating the Pareto optimal set. *Engineering Optimization*, **7**, 319–336.

Balachandran, M. and Gero J.S. (1985). The noninferior set estimation method for three objective problems. *Engineering Optimization*, **9**, 77–88.

Ballestero, E. and Romero, C. (1998). *Multiple Criteria Decision Making and its Applications to Economic Problems*. Kluwer Academic Publishers, Boston.

Barnett, D., Blake B. and McCarl, B.A. (1982). Goal programming via multidimensional scaling applied to Senegalese subsistence farms. *American Journal of Agricultural Economics*, **64**, 720–727.

Barry, P.J. Editor (1984). *Risk Management in Agriculture*. The Iowa State University Press, Iowa.

Bartlett, E.T. and Clawson, W.J. (1978). Profit, meat production or efficient use of energy in ranching. *Journal of Animal Science*, **46**, 812–818.

Bazaraa, M.S. and Bouzaher, A. (1981). A linear goal programming model for developing economies with an illustration from the agricultural sector in Egypt. *Management Science*, **27**, 396–413.

Belenson, S.M. and Kapur, K.C. (1973). An algorithm for solving multicriterion linear programming problems with examples. *Operational Research Quarterly*, **24**, 65–77.

Benayoun, R.J., de Montgolfier, J.T., Tergny and Laritchev, O. (1971). Linear programming with multiple objective functions: Step Method (STEM). *Mathematical Programming*, **1**, 366–375.

Berbel, J., Gallego, J. and Sagues, H. (1991). Marketing goals vs. business profitability: an interactive multiple criteria decision-making approach. *Agribusiness*, **7**, 537–549.

Blasco, F., Cuchillo-Ibáñez E., Morón M. A. and Romero, C. (1999).On the monotonocity of the compromise set in multicriteria problems. *Journal of Optimization Theory and Applications*, **102**, 69–82.

Buchanan, J.T. (1986). Multiple objective mathematical programming: a review. *New Zealand Operational Research*, **14**, 1–27.

Can, E.K. and Houck, M.H. (1984). Real-time reservoir operations by goal programming. *Journal of Water Resource Planning and Management*, **11**, 297–309.

Cary, J.W. and Holmes, W.E. (1982). Relationships among farmers' goals and farm adjustment strategies: some empirics of a multidimensional approach. *Australian Journal of Agricultural Economics*, **26**, 114–130.

Chankong, V. and Haimes, Y. (1983). *Multiobjective Decision Making: Theory and Methodology*. North-Holland, New York.

Charnes, A. and Cooper, W. (1959). Chance constrained programming. *Management Science*, **6**, 73–79.

Charnes, A. and Cooper, W.W. (1961). *Management Models and Industrial Applications of Linear Programming*, **1**, John Wiley and Sons, New York.

Charnes, A. and Cooper, W.W. (1975). Goal programming and constrained regression – A comment. *Omega*, **3**, 403–409.

Charnes, A. and Cooper, W.W. (1977). Goal programming and multiple objective optimization, part I. *European Journal of Operational Research*, **1**, 39–54.

Charnes, A., Cooper, W.W. and Ferguson, R. (1955). Optimal estimation of executive compensation by linear programming. *Management Science*, **1**, 138–151.

Choo, E.U. and Atkins, D.R.O. (1980). An interactive algorithm for multicriteria programming. *Computers and Operations Research*, **7**, 81–87.

Cochrane, J.L. and Zeleny, M. Editors, (1973). *Multiple Criteria Decision Making*. University of South Carolina Press, Columbia.

Cohon, J.L. (1978). *Multiobjective Programming and Planning*. Academic Press, New York.

Cohon, J.L. and Marks, D.H. (1975). A review and evaluation of multiobjective programming techniques. *Water Resources Research*, 11, 208–220.

Cohon, J.L., Church, R.L. and Sheer, D.P. (1979). Generating multiobjective trade-offs: an algorithm for bicriterion problems. *Water Resources Research*, 15, 1001–1010.

Colson, G. and Zeleny, M. (1980a). Multicriterion concept of risk under incomplete information. *Computers and Operations Research*, 7, 125–143.

Colson, G. and Zeleny, M. (1980b). *Uncertain Prospects Ranking and Portfolio Analysis Under the Conditions of Partial Information*. Mathematical Systems in Economics, No 44, Verlag Anton Hain, Meisenheim.

Cook, W.D. (1976). Zero-sum games with multiple goals. *Naval Research Logistics Quarterly*, 23, 615–622.

Coombs, C.H. (1958). On the use of inconsistency of preferences in psychological measurement. *Journal of Experimental Psychology*, 55, 1–7.

Crabtree, J.R. (1982). Interactive formulation system for cattle diets. *Agricultural Systems*, 8, 291–308.

De Wit, C. T., van Keulen, H., Seligman, N. G. and Spharim, I. (1988). Application of interactive goal programming techniques for analysis and planning of regional agricultural development. *Agricultural Systems*, 26, 211–230.

Debreu, G. (1959). *Theory of Value*. Cowles Foundation monograph 17, John Wiley and Sons, New York.

Díaz-Balteiro, L. and Romero, C. (1998). Modeling timber harvest scheduling problems with multiple criteria: an application in Spain. *Forest Science*, 44, 47–57.

Dillon J.L. (1962). Applications of game theory in agricultural economics: review and requiem. *Australian Journal of Agricultural Economics*, 6, 20–35.

Dinkelbach, W. and Iserman, H. (1980). Resource allocation of an academic department in the presence of multiple criteria – Some experience with a modified STEM-method. *Computers and Operations Research*, 7, 99–106.

Drynan, R.G. (1985). Goal programming and multiple criteria decision making in farm planning: and expository analysis – A comment. *Journal of Agricultural Economics*, 36, 421–423.

Duckstein, L. and Opricovic, S. (1980). Multiobjective optimization in river basin development. *Water Resources Research*, 16, 14–20.

Dyer, J.S. (1972). Interactive goal programming. *Management Science*, 19, 62–70.

Ecker, J.G., Hegner, N.S. and Kouada, I.A. (1980). Generating all maximal efficient faces for multiple objective linear programs. *Journal of Optimization Theory and Application,* **30**, 353–381.

Ehrgott, M. (2000). *Multicriteria Optimization.* Springer-Verlag, Berlin.

Ehrgott, M. and Gandibleux, X. Editors, (2002). *Multicriteria Optimization: State of the Art Annotated Bibliographic Survey.* Kluwer Academic Publishers, Boston

Eilon, S. (1972). Goals and constraints in decision-making. *Operational Research Quarterly,* **23**, 3–15.

Evans, G.W. (1984). An overview of techniques for solving multiobjective mathematical programs. *Management Science,* **30**, 1268–1282.

Evans, J.P. and Steuer, R.E. (1973). A revised simplex method for linear multiple objective programming. *Mathematical Programming,* **5**, 54–72.

Fichefet, J. (1976). GPSTEM: An interactive multiobjective optimization method. In: *Progress in Operations Research,* **1**, Prekopa, A. (Ed.), North-Holland, Amsterdam, 317–332.

Flavell, R.B. (1976). A new goal programming formulation. *Omega,* **4**, 731–732.

Flinn, J.C., Jayasuriya, S. and Knight, C.G. (1980). Incorporating multiple objectives in planning models of low-resource farms. *Australian Journal of Agricultural Economics,* **24**, 35–45

Forman, E. H. and Gass, S. I. (2001). The Analytic Hierachy Process-an exposition. *Operations Research,* **49**, 469–486.

France, J. and Thornley, J.H.M. (1984). *Mathematical Models in Agriculture.* Butterworth, London.

Franz, L. and Lee, S. (1981). A goal programming based interactive decision support system. In: *Organizations: Multiple Agents with Multiple Criteria,* Morse, J.N. (Ed.), Springer-Verlag, Berlin, 110–115.

Freimer, M. and Yu, P L. (1976). Some new results on compromise solutions for group decision making. *Management Science,* **22**, 688–693.

French, S. (1984). Interactive multi-objective programming: its aims, applications and demands. *Journal of the Operational Research Society,* **35**, 827–834.

Freund, R.J. (1956). The introduction of risk into a programming model. *Econometrica,* **24**, 253–264.

Friedman, M. (1962). *Price Theory: a Provisional Text.* Aldine, Chicago.

Gabbani, D. and Magazine, M. (1986). An interactive heuristic approach for multi-objective integer-programming problems. *Journal of the Operational Research Society,* **37**, 285–291.

Gal, T. (1986). On efficient sets in vector maximum problems – a brief survey. *European Journal of Operational Research*, **24**, 253–264.

Gardiner, L. R. and Steuer, R. E. (1994). Unified interactive multiple objective programming. *European Journal of Operational Research*, **74**, 391–406.

Garrod, N.W. and Moores, B. (1978). An implicit enumeration algorithm for solving zero-one goal programming problems. *Omega*, **6**, 374–377.

Gasson, R. (1973). Goals and values of farmers. *Journal of Agricultural Economics*, **24**, 521–537.

Gearhart, W.B. (1979). Compromise solutions and estimation of the noninferior set. *Journal of Optimization Theory and Applications*, **28**, 29–47.

Gearhart, W.B. (1984). Analysis of compromise programming. In: *MCDM: Past Decade and Future Trends*, Zeleny, M. (Ed.), JAI Press Inc., Connecticut, 85–100.

Geoffrion, A.M. (1968). Proper efficiency and the theory of vector maximization. *Journal of Mathematical Analysis and Applications*, **22**, 618–630.

Geoffrion, A.M., Dyer, J.S. and Feinberg, A. (1972). An interactive approach for multi-criterion optimization, with an application to the operation of an academic department. *Management Science*, **19**, 357–368.

Gershon, M. and Duckstein, L. (1983). Multiobjective approaches to river basin planning. *Journal of Water Resources Planning and Management Division*, ASCE, **109**, 13–28.

Goicochea, A., Hansen, D.R. and Duckstein, L. (1982). *Multiobjective Decision Analysis with Engineering and Business Applications*. John Wiley and Sons, New York.

Grauer, M. and Wierzbicki, A.P. (Eds.) (1984). *Interactive Decision Analysis*. Springer-Verlag, Berlin.

Greis, N.P., Wood, E.F. and Steuer, R.E. (1983). Multicriteria analysis of water allocation in a river basin: the Tchebycheff approach. *Water Resources Research*, **19**, 865–875.

Hafkamp, W. and Nijkamp, P. (1983). Conflict analysis and compromise strategies in integrated spatial systems. *Regional Science and Urban Economics*, **13**, 115–140.

Haimes, Y.Y. and Hall, W.A. (1974). Multiobjectives in water resources systems analysis: the Surrogate Worth Trade-off Method. *Water Resources Research*, **10**, 615–624.

Hannan, E.L. (1980). Nondominance in goal programming. INFOR. (*Canadian Journal of Research and Information Processing*), **18**, 300–309.

Hannan, E.L. (1982). Reformulating zero-sum games with multiple goals. *Naval Research Logistics Quarterly*, **29**, 113–118.

Hannan, E.L. (1984). Goal programming: methodological advances in 1973–1982 and prospects for the future. In: *MCDM: Past Decade and Future Trends*, Zeleny, M. (Ed.), JAI Press, Connecticut, 117–151.

Hannan, E.L. (1985). An assessment of some criticism of goal programming. *Computers and Operations Research*, **12**, 525–541.

Hardaker, J. B., Huirne, R. B. M. and Anderson, J. R. (1997). *Coping with Risk in Agriculture*. CAB International, Wallingford.

Harper, W.H. and Eastman, C.F. (1980). An evaluation of goal hierarchies for small farm operators. *American Journal of Agricultural Economics*, **62**, 742–747.

Harrald, J., Leotta, J., Wallace, W.A. and Wendell, R.E. (1978). A note on the limitations of goal programming as observed in resource allocation for marine environmental protection. *Naval Research Logistics Quarterly*, **25**, 733–739.

Hayashi, K. (2000). Multicriteria analysis for agricultural resource management: a critical survey and future perspectives. *European Journal of Operational Research*, **122**, 486–500.

Hazell, P.B.R. (1970). Game theory – An extension of its application to farm planning under uncertainty. *Journal of Agricultural Economics*, **21**, 239–252.

Hazell, P.B.R. (1971). A linear alternative to quadratic and semivariance programming for farm planning under uncertainty. *American Journal of Agricultural Economics*, **53**, 53–62.

Hitchens, M.T., Thampapillai, D.J. and Sinden, J.A. (1978). The opportunity cost criterion for land allocation. *Review of Marketing and Agricultural Economics*, **46**, 275–293.

Hwang, C.L., Masud, A.S.N., Paidy, S.R. and Yoon, K. (1979). *Multiple Objective Decision Making – Methods and Applications: a State-of-the-Art Survey*. Springer-Verlag, Berlin.

Ignizio, J.P. (1976). *Goal Programming and Extensions*. Lexington Books, Massachusetts.

Ignizio, J.P. (1978). A review of goal programming: a tool for multiobjective analysis. *Journal of the Operational Research Society*. **27**, 1109–1119.

Ignizio, J.P. (1981). The determination of a subset of efficient solutions via goal programming. *Computers and Operations Research*, **8**, 9–16.

Ignizio, J.P. (1982). *Linear Programming in Single and Multiple-objective Systems*. Prentice-Hall, New Jersey.

Ignizio, J.P. (1983). Generalized goal programming. An over-view. *Computer and Operations Research*, **10**, 277–289.

Ignizio, J.P. (1985). *Introduction to Linear Goal Programming*. Sage Publications, Beverley Hills, California.

Ignizio, J.P. and Perlis, J.H. (1979). Sequential linear goal programming. *Computers and Operations Research*, **6**, 141–145.

Ijiri, Y. (1965). *Management Goals and Accounting for Control.* North-Holland, Amsterdam.

Isermann, H. (1977). The enumeration of the set of all efficient solutions for a linear multiple objective program. *Operational Research Quarterly*, **28**, 711–725.

Jones, D. F. and Tamiz, M. (1995). Expending the flexibility of goal programming via preference modelling techniques. *Omega*, **23**, 41–48-

Jones, D. F., Tamiz, M. and Mirrazavi, S. K. (1998). Intelligent solution and analysis of goal programmes: the GPSYS system. *Decision Support Systems*, **23**,329–332.

Kawaguchi, T. and Maruyama, Y. (1972). Generalized constrained games in farm planning. *American Journal of Agricultural Economics*, **54**, 591–602.

Keen, P.G.W. (1977). The evolving concept of optimality, In: *Multiple Criteria Decision Making*, Starr, M.K. and Zeleny, M. (Eds.). North-Holland Publishing, Amsterdam, 31–58.

Keeney, R.L. and Raiffa, H. (1976). Decision with multiple objectives: *Preferences and Value Trade-Offs.* John Wiley and Sons, New York.

Keown, A.J. (1978). A chance-constrained goal programming model for bank liquidity management. *Decision Science*, **9**, 93–106.

Keown, A.J. and Martin, J.D. (1977). A chance-constrained goal programming model for working capital management. *The Engineering Economist*, **22**, 152–174.

Knight, F.H. (1921). *Risk, Uncertainty and Profit.* Houghton Mifflin Company, New York.

Koopmans, T.C. (1951). Analysis of production as an efficient combination of activities. In: *Activity Analysis of Production and Allocation*, Koopmans, T.C. (Ed.), John Wiley and Sons, New York, 33–97.

Korhonen, P. and Laakso, J. (1986). A visual interactive method for solving the multiple criteria problem. *European Journal of Operational Research*, **24**, 277–287.

Kornbluth, J.S.H. (1973). A survey of goal programming, *Omega*, **1**, 193–205.

Kornbluth, J.S.H. (1984). Fractional programming in MCDM. In: *MCDM: Past Decade and Future Trends*, Zeleny, M. (Ed.), JAI Press Inc., Connecticut 153–167.

Kornbluth, J.S.H. and Steuer, R.E. (1981). Multiple objective linear fractional programming. *Management Science*, **27**, 1024–1039.

Kuhn, H.W. and Tucker, A.W. (1951). Nonlinear programming. In: *Proceedings of the Second Berkeley Symposium on Mathematical Statistics and Probability*, Neyman, J. (Ed.), University of California Press, Berkeley, 481–491.

Kvanli, A.H. (1980). Financial planning using goal programming. *Omega*, **8**, 207–218.

Lara, P. and Romero, C. (1992). An interactive multigoal programming model for determining livestock rations: an application to dairy cows in Andalusia, Spain. *Journal of the Operational Reserach Society*, **43**, 945–953.

Lee, S.M. (1972). *Goal Programming for Decision Analysis*. Auerbach Publishers, Philadelphia.

Lee, S.M. and Morris, R. (1977). Integer goal programming methods, In: *Multiple Criteria Decision Making*, Starr, N. and Zeleny, M. (Eds.), North-Holland, Amsterdam, 272–289.

Lin, W.T. (1980). A survey of goal programming applications. *Omega*, **18**, 115–117.

Liu, G. and Davis, L.S. (1995). Interactive resolution of multi-objective forest planning problems with shadow price and parametric analysis. *Forest Science*, **41**, 452–469.

Marcotte, O. and Soland, R.M. (1986). An interactive branch-and-bound algorithm for multiple criteria optimization. *Management Science*, **32**, 61–75.

Marglin, S.A. (1967). *Public Investment Criteria*. M.I.T. Press, Cambridge, Massachusetts.

Markowitz, H. (1952). Portfolia Selection, *Journal of Finance*, **7**, 77–91.

Marten, G.G. and Sancholuz, L.A. (1982). Ecological land-use planning and carring capacity evaluation in the Jalapa region (Veracruz, Mexico). *Agro-Ecosystems*, **8**, 83–124.

Masud, A.S. and Hwang, C.L. (1981). Interactive sequential goal programming. *Journal of the Operational Research Society*, **32**, 391–400.

Mathiesen, L. (1981). A multi-criteria model for assessing industrial structure in the Norwegian fish-meal industry. In: *Applied Operations Research in Fishing*, Haley, K.B. (Ed.), Plenum Press, New York, 281–294.

McCarl, B.A. and Blake, B.F. (1983). Goal programming via multi-dimensional scaling applied to Senegalese subsistence farming: Reply. *American Journal of Agricultural Economics*, **65**, 832–833.

McInerney, J.P. (1967). 'Maximing programming' – an approach to farm planning under uncertainty, *Journal of Agricultural Economics*, **18**, 279–289.

McInerney, J.P. (1969). Linear programming and game theory models – some extensions. *Journal of Agricultural Economics*, **20**, 269–278.

Michalowski, W. (1981). The choice of final compromise solution in multiple criteria linear programming problem, In: *Organization: Multiple Agents with Multiple Criteria*, Morse, J.N. (Ed.), Springer-Verlag, Berlin, 233–238.

Michalowski, W. (1985). An experiment with Ziont-Wallenius and Steuer interactive programming models. In: *Decision Making with Multiple Objectives*, Haimes, Y.Y. and Chankong, V. (Eds.), Springer-Verlag, Berlin, 424–429.

Michalowski, W. and Piotrowski, A. (1983). Solving a multi-objective planning problem by an interactive procedure. In: *Multi-objective Decision Making*, French, S., Hartley, R., Thomas, L.C. and White, D.J. (Eds.), Academic Press, London, 235–249.

Michalowski, W. and Zolkiewski, Z. (1983). An interactive approach to the solution of a linear production planning problem with multiple objectives. In: *Essays and Surveys on Multiple Criteria Decision Making*, Hansen, P. (Ed.), Springer-Verlag., Berlin, 260–268.

Miettinen, K. M. (1999). *Nonlinear Multiobjective Optimization*. Kluwer Academic Publishers, Boston.

Minguez, M.I., Romero, C. and Domingo, J., (1988). Determining fertilizer combination through goal programming with penalty functions. An application to sugar beet production in Spain. *Journal of the Operational Research Society*, **39**, 61–70.

Monarchi, D.E., Weber, J.E. and Duckstein, L. (1976). An interactive multiple objective decision-making aid using nonlinear goal programming. In: *Multiple Criteria Decision Making*. Kyoto 1975. Zeleny, M. (Ed.), Springer-Verlag, Berlin, 235–253.

Nijkamp, P. and Spronk, J. (1978). Goal programming for decision making: an overview and a discussion. *Ricerca Operativa*, 33–49.

Nijkamp, P. and Spronk, J. (1980). Interactive multiple goal programming: an evaluation and some results. In: *Multiple Criteria Decision Making Theory and Application*, Fandel, G. and Gal, T. (Eds.), Springer-Verlag, Berlin, 278–293.

Olson, D.L. (1984). Comparison for four goal programming algorithms. *Journal of the Operational Research Society*, **35**, 347–354.

Olson; D. L. (1992). Review of empirical studies in multiobjective mathematical programming: subject reflection of nonlinear utility and learning. *Decision Sciences*, **23**, 1–20.

Patrick, G.F. and Kliebenstein, J.B. (1980). *Multiple Goals in Farm Firm Decision-making: a Social Science Perspective*. Department of Agricultural Economics, Agricultural Experimental Station, Purdue University.

Philip, J. (1972). Algorithms for the vector maximization problem. *Mathematical Programming*, 2, 207–229.

Ramesh, R., Zionts, S. and Karwan, M.H. (1986). A class of practical interactive branch and bound algorithms for multicriteria integer programming. *European Journal of Operational Research*, **26**, 161–172.

Rehman, T. and Romero, C. (1984). Multiple-criteria decision-making techniques and their role in livestock ration formulation. *Agricultural Systems*, **15**, 23–49.

Rehman, T. and Romero, C. (1987a). Multiobjective and goal programming techniques for solving agricultural planning problems. In: *Agriculture and Economic Instability*, Bellamy, M. and Greenshields, B. (Eds.), Gower Publishing Company, Aldershot, 355–359.

Rehman, T. and Romero, C. (1987b). Goal programming with penalty functions and livestock ration formulation. *Agricultural Systems*, **23**, 117–132.

Rehman, T. and Romero, C. (1993). The application of the MCDM paradigm to the management of agricultural systems: some basic considerations. *Agricultural Systems*, **41**, 239–255.

Ringuest, J. L. (1992). *Multiobjective Optimization: Behavioral and Computational Considerations*. Kluwer Academic Publisheres, Boston.

Romero, C. (1984). A note – Effects of five sided penalty functions in goal programming. *Omega*, **12**, 333.

Romero, C. (1986). A survey of generalized goal programming (1970–1982). European *Journal of Operational Research*, **25**, 183–191.

Romero, C. (1991). *Handbook of Critical Issues in Goal Programming*. Pergamon Press, Oxford.

Romero, C. (2000). Risk programming for agricultural resource allocation. A multidimensional risk approach. *Annals of Operations Research*, **94**, 57–68.

Romero, C. (2001). Extended lexicographic goal programming: a unifying approach. *Omega*, **29**, 63–71.

Romero, C., Amador, F. and Barco, A. (1987). Multiple objectives in agricultural planning: a compromise programming application. *American Journal of Agricultural Economics*, **69**, 78–86.

Romero, C. and Rehman, T. (1983). Goal programming via multidimensional scaling applied to Senegalese subsistence farming: comment. *American Journal of Agricultural Economics*, **65**, 829–831.

Romero, C. and Rehman, T. (1984). Goal programming and multiple criteria decision-making in farm planning: an expository analysis. *Journal of Agricultural Economics*, **35**, 177–190.

Romero, C. and Rehman, T. (1985a). Goal programming and multiple criteria decision making in farm planning: some extensions. *Journal of Agricultural Economics*, **36**, 171–185.

Romero, C. and Rehman, T. (1985b). Goal programming and multiple criteria decision-making in farm planning: an expository analysis – a reply. *Journal of Agricultural Economics*, **36**, 425–427.

Romero, C. and Rehman, T. (1987). Natural resource management and the use of multiple criteria decision making techniques: a review. *European Review of Agricultural Economics*, **14**, 61–89.

Romero, C., Rehman, T. and Domingo, J. (1988). Compromise-risk programming for agricultural resource allocation: an illustration. *Journal of Agricultural Economics*, **39**, 271–276.

Roy, B. (1971). Problems and methods with multple objective functions. *Mathematical Programming*, **1**, 239–266.

Roy, B. and Vincke, P. (1981). Multicriteria analysis: survey and new directions. *European Journal of Operational Research*, **8**, 207–218.

Saaty, T. L (1977). A scaling method for priorities in hierarchical structures. *Journal of mathematical psychology*, **15**, 234–281.

Saaty, T. L. (1980). *The Analytic Hierarchy Process: Planning, Priority Setting, and Resource Allocation*. McGraw-Hill, New York.

Sakawa, M. (1982). Interactive multiobjective decision making by the sequential proxy optimization technique: SPOT. *European Journal of Operational Research*, **9**, 386–396.

Sakawa, M. (1984). Interactive fuzzy goal programming for multiobjective nonlinear problems and its application to water quality management. *Control and Cybernetics*, **13**, 217–228.

Samblanckx, S., Depraetere, P. and Muller, H. (1982). Critical considerations concerning the multicriteria analysis by the method of Zionts and Wallenius. *European Journal of Operational Research*, **10**, 70–76.

Schniederjans, M.J. (1984). *Linear Goal Programming*. Petrocelli Books, New Jersey.

Schniederjans, M. J. (1995). *Goal Programming. Methodology and Applications*. Kluwer Academic Publishers, Boston.

Schniederjans, M.J. and Kwak, N.K. (1982). An alternative solution method for goal programming problem: a tutorial. *Journal of the Operational Research Society*, **33**, 247–251.

Slowinski, R. and Warczynski, J. (1984). Application of the Ellipsoid Method in an interactive procedure for multicriteria linear programming. *Zeitschrift fur Operations Research*, **28**, 89–100.

Smith, D. and Capstick, D. (1976). Establishing priorities among multiple management goals. *Southern Journal of Agricultural Economics*, **2**, 37–43.

Spronk, J. (1981). *Interactive Multiple Goal Programming*. Martinus Nijhoff Publishing, Boston.

Spronk, J. and Telgen, J. (1981). An ellipsoidal interactive multiple goal programming method. In: *Organizations: Multiple Agents with Multiple Criteria*, Morse, J.N. (Ed.), Springer-Verlag, Berlin, 380–387.

Spronk, J. and Veeneklaas, F. (1983). A feasibility study of economic and environmental scenarios by means of interactive multiple goal programming. *Regional Science and Urban Economics*, **13**, 141–160.

Spronk, J. and Zambruno, G. (1985). Interactive multiple goal programming for bank portfolio selection. In: *Multiple Criteria Decision Methods and Applications*, Fandel, G. and Spronk, J. (Eds.), Springer-Verlag, Berlin, 289–306.

Stadler, W. (1984) A comprehensive bibliography on multi-criteria decision making. In: *MCDM Past Decade and Future Trends*, Zeleny, M. (Ed.), JAI Press Inc., Connecticut, 223–328.

Stancu-Minasian, I. M. (1997). *Fractional Programming. Theory, Methods and Applications.* Kuwer Academic Publishers, Dordrecht.

Steuer, R.E. (1976). Multiple objective linear programming with interval criterion weights. *Management Science*, **23**, 305–316.

Steuer, R.E. (1977). An interactive multiple objective linear programming procedure. In: *Multiple Criteria Decision Making*, Starr, M.K. and Zeleny, M. (Eds.), North-Holland, Amsterdam.

Steuer, R.E. (1986). *Multiple Criteria Optimization: Theory, Computation and Application.* John Wiley and Sons, New York.

Steuer, R. E. (1995). *Manual for the ADBASE Multiple Objective Linear Programming Package.* Faculty of Management Science, University of Georgia, Athens.

Steuer, R.E. and Choo, E.U. (1983). An interactive weighted Tchebycheff procedure for multiple objective programming. *Mathematical Programming*, **26**, 326–344.

Steuer, R. E., Gardiner, L. R. and Gray, J. (1996). A bibliographic survey of the activities and international nature of multiple criteria decision making. *Journal of Multi-Criteria Decision Analysis*, **5**, 195–217.

Steuer, R.E. and Harris, F.W. (1980). Intra-set point generation and filtering in decision and criterion space. *Computers and Operations Research*, **7**, 41–53.

Stewart, T.J. (1984). Inferring preferences in multiple criteria decision analysis using a logistic regression model. *Management Science*, **30**, 1067–1077.

Stewart, T.J. (1986). A combined logistic regression and Zionts-Wallenius methodology for multiple criteria linear programming. *European Journal of Operational Research*, **24**, 295–304.

Stewart, T.J. (1988). Experience with prototype multicriteria decision support systems for pelagic fish quota determinations. *Naval Research Logistics*, **35**, 719–731.

Szidarovszky, F., Gershon, M.E. and Duckstein, L. (1986). *Techniques for Multiobjective Decision Making in Systems Management*. Elsevier Science, Amsterdam.

Tabucanon, M.T. and Mukyangkoon, S. (1985). Multi-objective microcomputer-based interactive production planning. *International Journal of Production Research*, **23**, 1001–1023.

Tamiz, M. and Jones, D. F. (1996). Goal programming and Pareto efficiency. *Journal of Information and Optimization Sciences*, **17**, 291–307.

Tamiz, M., Jones, D. F. and Romero, C. (1998). Goal programming for decision making: an overview of the current state-of-the-art. *European Journal of Operational Research*, **111**, 569–581.

Tauer, L.W. (1983). Target MOTAD. *American Journal of Agricultural Economics*, **65**, 606–610.

Teghem, J. and Kunsch, P.L. (1985). Application of multi-objective stochastic linear programming to power system planning. *Engineering Costs and Production Economics*, **9**, 83–89.

Teghem, J., Dufrane, D., Thauvoye, M. and Kunsch, P. (1986). STRANGE: an interactive method for multi-objective linear programming under uncertainty. *European Journal of Operational Research*, **26**, 65–82.

Thampapillai, D.J. (1978). Methods of multiple objectives planning: A review. World *Agricultural Economics and Rural Sociology Abstracts*, **20**, 803–813.

Thampapillai, D.J. and Sinden, J.A. (1979). Trade-offs for multiple objective planning through linear programming. *Water Resources Research*, 15, 1028–1033.

Vedula, S. and Rogers, P. (1981). Multiple analysis of irrigation planning in river basin development. *Water Resources Research*, **17**, 1304–1310.

Vincke, Ph. (1986). Analysis of multicriteria decision aid in Europe. *European Journal of Operational Research*, **25**, 160–168.

von Neumann, J. and Morgenstern, O. (1944). *Theory of Games and Economic Behaviour*. Princeton University Press, Princeton, New Jersey.

Walker, H.D. (1985). An alternative approach to goal programming. *Canadian Journal of Forest Research*, **15**, 319–325.

Wallenius, J. (1975a). *Interactive Multiple Criteria Decisions Method: an Investigation and an Approach*. The Helsinki School of Economics, Helsinki.

Wallenius, J. (1975b). Comparative evaluation of some interactive approaches to multicriterion optimization. *Management Science*, **21**, 1387–1396.

Wallenius, J. and Zionts, S. (1977). A research project on multicriterion decision making. In: *Conflicting Objectives in Decisions*, Bell, D.E., Keeney, R.L. and Raiffa, H. (Eds.), John Wiley and Sons, New York.

Walsh, V.C. (1970). *Introduction to Contemporary Microeconomics*. McGraw-Hill, New York.

Weistroffer, H.R. (1983). An interactive goal programming method for nonlinear multiple-criteria decision-making problems. *Computers and Operations Research*, **10**, 311–320.

Weistroffer, H.R. (1984). A combined over -and under-achievement programming approach to multiple objectives decision-making. *Large Scale Systems*, **7**, 47–58.

Wheeler, B.M. and Russell, J.R.M. (1977), Goal programming and agricultural planning. *Operational Research Quarterly*, **28**, 21–32.

White, D.J. (1980). Multi-objective interactive programming. *Journal of the Operational Research Society*, **31**, 517–523.

White, D.J. (1983). The foundations of multi-objective interactive programming – some questions. In: *Essays and Surveys on Multiple Criteria Decisions Making*, Hansen, P. (Ed.), Springer-Verlag, Berlin, 406–414.

Wierzbicki, A.P. (1982). A mathematical basis for satisficing decision making. *Mathematical Modelling*, **3**, 391–405.

Willis, C.E. and Perlack, R.D. (1980). A comparison of generating techniques and goal programming for public investment, multiple objective decision making. *American Journal of Agricultural Economics*, **62**, 66–74.

Yilmaz, M. (1984). A theory of the displaced ideal with decisions under uncertainty. In: *MCDM, Past Decade and Future Trends*, Zeleny, M. (Ed.), JAI Press Inc., Connecticut, 101–116.

Yu, P.L. (1973). A class of solutions for group decision problems. *Management Science*, **19**, 936–946.

Yu, P.L. (1985). *Multiple Criteria Decision Making: Concepts, Techniques and Extensions*. Plenum, New York.

Yu, P.L. and Leitman, G. (1974). Compromise solutions, domination structures, and Salukvadze's solution. *Journal of Optimization Theory and Applications*, **13**, 362–378.

Yu, P.L. and Zeleny, M. (1975). The set of all nondominated solutions in linear cases and a multicriteria simplex method. *Journal of Mathematical Analysis and Applications*, **49**, 430–468.

Zadeh, L.A. (1963). Optimality and non-scalar-valued performance criteria. *IEEE Transactions on Automatic Control*, **AC-8**, 59–60.

Zanakis, S.H. and Gupta, S.K. (1985). A categorized bibliographic survey of goal programming. *Omega*, **13**, 211–222.

Zeleny, M. (1973). Compromise programming. In: *Multiple Criteria Decision Making*, Cochrane, J.L. and Zeleny, M. (Eds.), University of South Carolina Press, Columbia, 262–301.

Zeleny, M. (1974). *Linear Multiobjective Programming*. Springer-Verlag, Berlin.

Zeleny, M. (1976a). Multicriteria simplex method: a FORTRAN routine. In: *Multiple Criteria Decision Making*. Kyoto 1975, Zeleny, M. (Ed.), Springer-Verlag, Berlin, 323–345.

Zeleny, M. (1976b). Games with multiple payoffs. *International Journal of Game Theory*, **4**, 179–191.

Zeleny, M. (1981). The pros and cons of goal programming. *Computers and Operations Research*, **8**, 357–359.

Zeleny, M. (1982). *Multiple Criteria Decision Making*. McGraw-Hill, New York.

Zeleny, M. and Cochrane, J.L. (1973). A priori and a posteriori goals in macroeconomic policy making. In: *Multiple Criteria Decision Making*, Cochrane, J.L. and Zeleny, M. (Eds.), University of South Carolina Press, South Carolina, 373–391.

Zionts, S. (1980). Methods for solving management problems involving multiple objectives. In: *Multiple Criteria Decision Making, Theory and Applications*, Fandel, G. and Gal, T. (Eds.), Springer-Verlag, Berlin.

Zionts, S. (1981). A multiple criteria method for choosing among discrete alternatives. *European Journal of Operational Research*, **7**, 143–147.

Zionts, S. (1983). A report on a project on multiple criteria decision making, 1982. In: *Essays and Surveys on Multiple Criteria Decision Making*, Hansen, P. (Ed.), Springer-Verlag, Berlin, 416–430.

Zionts, S. and Deshpande, D. (1978). A time sharing computer programming application of a multiple criteria decision method to energy planning – a progress report. In: *Multiple Criteria Problem Solving*, Zionts, S. (Ed.), Springer-Verlag, Berlin, 549–560.

Zionts, S. and Deshpande, D. (1981). Energy planning using a multiple criteria decision method. In: *Multiple Criteria Analysis*, Nijkamp, P. and Spronk, J. (Eds.), Gower Publishing Company, Aldershot, 153–162.

Zionts, S. and Wallenius, J. (1976). An interactive programming method for solving the multiple criteria problem. *Management Science*, **22**, 652–663.

Zionts, S. and Wallenius, J. (1980). Identifying efficient vectors: some theory and computational results. *Operations Research*, **28**, 785–793.

Zionts, S. and Wallenius, J. (1983). An interactive multiple objective linear programming method for a class of underlying nonlinear utility functions. *Management Science,* **29**, 519–529.

Index

Absolute weights, see pre-emptive weights
Achievement function, 28, 30, 39
Agrarian reform, 123–124
Agrawal-Heady criterion, 111–112, 114, 117–118
Alternative optimal solutions, 33–36, 40–41, 53–54
Anti-ideal point, 52, 67, 82–83
Aspiration level, definition of, see target
Attributes, definition of, 15–16

Benefit criterion, see Agrawal-Heady criterion
Best-compromise solution, 10, 66–70, 110
Business profitability, 123

Choice, axiom of, 66
Compromise games, 116–119
Compromise programming
    case study, 123–134
    comparison with goal and multiobjective
        programming, 74–78
    concept, 10, 63
    continuous setting, 68–71
    discrete approximation, 66–68
Compromise-risk programming, 108–110
Compromise set, 70–74, 127–129, 132
Compromise strategy, 116
Constraints, difference from goals, 16–17

Decision variables space, 49
Deviational variables, definition of
    positive, 16–17, 26–27
    negative, 16–17, 26–27
Displaced ideal method, 71–74
Distance measures, 63–66

Euclidean, 64
    Chebysev, 66
Economic decisions versus technological
    problems, 4–7
Efficient solutions, 8, 17–18, 26, 47–50
Expectation criterion, 114, 117
Extreme efficient points, 48–50, 53–54, 59

Farm planning, 23–26, 125–126
Fertilizer combination
    problem situation, 163, 167

weighted goal programming with penalty
    functions, 171, 175
Filtering techniques, 60, 132

Game theory, 110–113
Games against nature, 110
Games with multiple goals, 113–115
Goal programming
    chance constrained, 42
    concept, 7–8, 19
    critical assessment, 32–41
    dominated solutions with, 40
    fractional, 41
    general aspects, 23–45
    integer, 37
    MINMAX, 41, 77
    nonlinear, 37
    with penalty functions, 44, 152, 157, 167
Goals, definition of, 16–17

Ideal point, 51–52, 63, 66–69, 71–74, 76–78, 82–83,
    85, 107–109, 126–128
Interactive programming
    assessment, 97–100
    concept, 10, 79–80
    structure, 80–81
Interactive multiple goal programming
    algorithm, 92–97
    assessment, 99–100
    flowchart, 98
Interior efficient points, 54, 59

Lexicographic goal programming
    algorithms, 33–37
    applications, 142–146
    concept, 23, 27–30
    graphical method, 31–33
    naïve prioritisation, 39–41
    sequential linear method, 33–36
Livestock ration formulation
    assessment, 160–161
    lexicographic goal programming, 142–146
    lexicographic goal programming
        with penalty functions, 157, 159–160

multiobjective programming
    problem situation, 146–147
weighted goal programming, 138–142
weighted goal programming with
    penalty functions, 152–157

Maximin criterion, see Wald criterion
Minimax regret criterion, see Savage criterion
Modified simplex, 36
MOTAD, 103–106, 110
Multiattribute utility theory, 19–20
Multigoal programming 58–59
Multiobjective programming
    applications, 123–133, 146–147
    concept, 19, 41–51
    constrained method, 52–53
    weighting method, 54–55
Multiobjective simplex method, 51, 61
Multiple goals in agriculture, 7–8
Multiple objectives in agriculture, 7–8

Nadir point, see anti-ideal point
Neoclassical theory, 71
Net present value, 24–26, 47–51
Nijkamp and Spronk, method of, see interactive
    multiple goal programming
NISE method, 55–59, 61
Non-dominated solutions, see efficient solutions

Objectives space, 49
Objectives, definition of, 15–16
Operational framework for decision-making, 3–4

Parametric games, 118
Pareto optimality, see efficient solutions
Partitioning algorithm, 36
Pay-off matrix
    application, 126
    in compromise-risk programming, 108
    in game theory, 110–111, 113
    in multiobjective programming, 51–52, 55, 126
    in the STEM method, 82, 84
Potency matrix, 94–96
Pre-emptive weights, 9, 23, 27–30
Prospect ranking vector, 119
Pruning of efficient sets, see filtering techniques

Relative weights, 23
Risk analysis, 103–106
Risk aversion, 110
Risk programming, method of Markowitz,
    103–105, 110

Satisficing, 44, 58
Savage criterion 110–112, 117–119
Seasonal labour, measure of, 125–133
STEM method
    algorithm, 81–85
    assessment, 97, 100
    flowchart 86

Target, definition of, 16
Target MOTAD, 104, 110
Trade-off curve, 49–50, 107, 127–128
Trade-offs
    concept, 18–19
    in goal programming, 39
    in multiobjective programming, 48–51
    in the interactive approach, 80–81
Transformation curve, see trade-off

Uncertainly, analysis of, 103–106

Value added, 5–7
Vector optimisation, see multiobjective
    programming

Wald criterion, 110–114, 116–119
Weighted goal programming
    applications, 138–142, 152–157, 163–167
    concept, 23
    general aspects, 37–38

Zionts and Wallenius, method of
    algorithm, 88
    assessment, 97, 99–100
    flowchart, 93

Printed and bound by CPI Group (UK) Ltd, Croydon, CR0 4YY

08/05/2025

01864927-0001